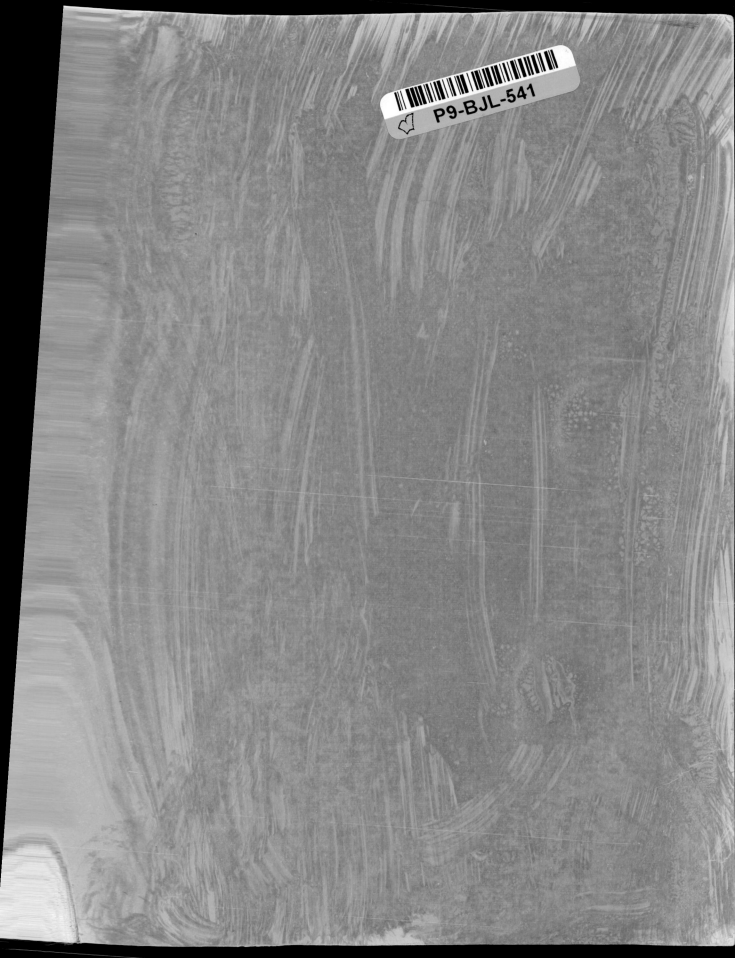

*INTERNATIONAL SERIES OF MONOGRAPHS*
*PURE AND APPLIED BIOLOGY*

Division: **PLANT PHYSIOLOGY**
General Editors: P. F. WAREING and A. Y. GALSTON

VOLUME 5

# THE GERMINATION OF SEEDS

## OTHER TITLES OF INTEREST

# The Germination of Seeds

## Second Edition

BY

## A. M. MAYER

*Professor of Botany, The Hebrew University of Jerusalem*

and

## A. POLJAKOFF-MAYBER

*Professor of Botany, The Hebrew University of Jerusalem*

## PERGAMON PRESS

OXFORD · NEW YORK · TORONTO
SYDNEY · PARIS · BRAUNSCHWEIG

| U.K. | Pergamon Press Ltd., Headington Hill Hall, Oxford OX3 0BW, England |
| U.S.A. | Pergamon Press Inc., Maxwell House, Fairview Park, Elmsford, New York 10523, U.S.A. |
| CANADA | Pergamon of Canada, Ltd., 207 Queen's Quay West, Toronto 1, Canada |
| AUSTRALIA | Pergamon Press (Aust.) Pty. Ltd., 19a Boundary Street, Rushcutters Bay, N.S.W. 2011, Australia |
| FRANCE | Pergamon Press SARL, 24 rue des Ecoles, 75240 Paris, Cedex 05, France |
| WEST GERMANY | Pergamon Press GMbH, 3300 Braunschweig, Postfach 2923, Burgplatz 1, West Germany |

First edition 1963

Second edition 1975

Library of Congress Cataloging in Publication Data

Mayer, A    M
The germination of seeds.

(International series of pure and applied biology:
Division, plant physiology; v. 5)
Includes bibliographies and indexes.
1. Germination. I. Poljakoff-Mayber, Alexandra,
1915–    joint author. II. Title.
QK740.M38 1975    581.3′3′3    75-12800
ISBN 0-08-018966-0
ISBN 0-08-018965-2 pbk.

*Printed in Great Britain by A. Wheaton & Co.*

# CONTENTS

# PREFACE TO THE SECOND EDITION

In preparing the new edition of this book we were faced with the difficulty of bringing it up to date, in view of the very large number of papers published in the last 12 years, and yet remaining within a reasonable framework as far as size is concerned. The original structure of this book has been retained, but the last chapter dealing with cryptobiotic states in general has been omitted. This helped to compensate partly for the increase in size of this book.

The material in all the chapters has been brought up to date. This will be marked especially in Chapters 4, 5 and 6 to which new subsections have been added, dealing with dormancy-inducing hormones, metabolism and especially protein and nucleic acid metabolism and information on the metabolic effects of growth promoters and inhibitors such as gibberellic acid, cytokinins and abscisic acid. A section on seed establishment has been added in Chapter 7.

In order to increase the usefulness of the book for those who wish to extend their reading the literature citations have been increased. We felt that the reader should be able to go as far as possible directly to the newer literature, which is dispersed among very many journals in different disciplines.

As in the first edition we have been selective in our choice of topics and the data cited and the selections made were necessarily very subjective. Obviously the field of germination has not been completely covered. Nevertheless, we hope that a reasonable balance has been maintained.

We are grateful to all those who have granted us permission to reproduce or otherwise utilize data from their published work.

Our thanks are due to especially Dr. D. E. Briggs and Academic Press for the use of Fig. 6.3; to Drs. Gutterman and Shain for the electron micrographs in Fig. 1.3i and Fig. 5.15b and Mrs. Marbach for that in Fig. 1.3j; and to Drs. D. E. Briggs, W. R. Briggs, M. J. Chrispeels, E. E. Dekker, L. S. Dure, M. M. Edwards, L. Fowden, R. H. Hageman, B. Juliano, D. L. Laidman, C. Kolloffel, A. Marcus, E. Marre, L. G. Paleg, P. Rollin and P. F. Wareing for material as cited in the text.

We hope that this new edition will encourage more biologists to enter the area of research on seed germination. It is obvious that very much is still to be learned on this subject. Seed germination is of great importance in biology, agriculture and food production for the human race. Its study still presents a great challenge to the inquiring mind.

*Jerusalem,*        A. M. M.
*November, 1974*        A. P. M.

# PREFACE TO THE FIRST EDITION

Although this volume on germination is one of a series of monographs on subjects in plant physiology, we have not attempted to cover the entire field of germination. The number of papers is vast and goes back 50 to 100 years. For this reason we have cited those papers which appeared to us as of importance. No doubt important papers have been omitted and differences of opinion are possible both as regards the selection and the arrangement of the material. An attempt has been made to treat the available information critically and to arrange it in an integrated form. At the same time it is impossible to avoid personal predilections and this has resulted in a more extensive treatment of those chapters dealing with subjects of special interest to the authors.

Our thanks are due to many scientists who have given us permission to use figures, tables and data from their published works, as follows: Professor P. Maheshwari and the McGraw-Hill Book Co. for Fig. 1.1, Professor W. Troll and the Gustav Fisher Verlag for Fig. 1.3(a) and (b). Professor R. M. McLean and Longmans Green and Co. Ltd. for Fig. 1.3(g), MacMillan and Co. Ltd. for Figs. 1.3(b), (f) and (h), and Fig. 1.4(a) from Hayward: *Structure of Economic Plants*, Dr. E. E. Conn and the *Journal of Biological Chemistry* for Table 5.9, as well as to Drs. Y. Oota, M. Yamada, J. E. Varner, Y. Tazakawa, H. Albaum, S. P. Spragg, E. W. Yemm, H. Halvorson, A. Marcus, S. B. Hendricks, H. B. Sifton, S. Isikawa, P. F. Wareing, M. Holden, S. Brohult and A. M. McLeod for the material as cited in the text.

Our thanks are also due to others who have helped us in the preparation of the material.

Special thanks are due to Professor A. Fahn for reading and commenting on Chapter 1, and to the late Professor Hestrin and Dr. A. Lees for reading and commenting on Chapter 8, which is an attempt to demonstrate that the behaviour of seeds in their natural environment is not an isolated phenomenon in biology.

Lastly we are greatly indebted to our editor, Professor P. F. Wareing, for his critical reading of the entire manuscript and for his many comments and suggestions, which helped to give the book its present form. Nevertheless, it goes without saying that the views expressed and the attitude taken are entirely our responsibility.

*Jerusalem*

A. M. M.
A. P. M.

Chapter 1

# THE STRUCTURE OF SEEDS AND SEEDLINGS

The term germination is used to refer to a fairly large number of processes, including the germination of seeds, and of spores of bacteria, fungi and ferns as well as the processes occurring in the pollen grain when the pollen tube is produced. Although all these are processes of germination, we will confine the use of the term germination to the seeds of higher plants, the Angiosperms. Extension beyond this would lead to discussions which are well outside the scope of this monograph.

The seed of Angiosperms is essentially simple in structure and develops from a fertilized ovule. It consists of an embryo surrounded by an envelope. The embryo is usually derived from the fusions of nuclei of the male and female gametes, i.e. it is the result of the fertilization of the egg cell in the embryo sac by one of the male nuclei from the pollen tube. The envelope or testa originates from the mother plant and normally develops from the integuments of the ovule. In addition Angiosperm seeds contain an endosperm. The endosperm may persist as a storage organ or it may degenerate, remaining as a rudimentary tissue particularly in those cases where the cotyledons serve as storage organs. The endosperm may become fused to the seed or fruit coat.

The endosperm is derived, like the embryo, from maternal and paternal genetic material. Its primary nucleus arises from the triple fusion of two of the polar nuclei with one of the sperm nuclei (Fig. 1.1). Frequently the endosperm tissue is triploid, but its ploidy can be very high, and differ in various regions of the endosperm. This polyploidy arises from secondary processes after division of the primary nucleus. The endosperm may remain coenocytic throughout or it may show cellular organization. In some cases the coenocytic endosperm remains completely liquid as in the case of *Cocos nucifera* (Bhatnagar and Johri, 1972). In some plants seeds arise by apomixis or nonsexual processes, the seed being formed usually from a diploid nucleus. Cases of polyembryony can arise either due to the division of the zygotic embryo, as in orchids, or due to secondary embryo development from the nucellar tissue, as in some citrus species. Various other aberrations in seed formation are also known.

In addition to these three basic structures, embryo, testa and endosperm, other tissues may occasionally participate in the seed's make-up, which is by no means constant. The testa is sometimes made up of tissues of the parent plant other than the integuments, e.g. the nucellus, the endosperm and sometimes even the chalaza. The testa itself varies greatly in form. It may be soft, gelatinous or hairy, although a hard testa is the form most commonly met.

The size and shape of seeds is extremely variable. It depends on the form of the ovary, the condition under which the parent plant is growing during the seed formation

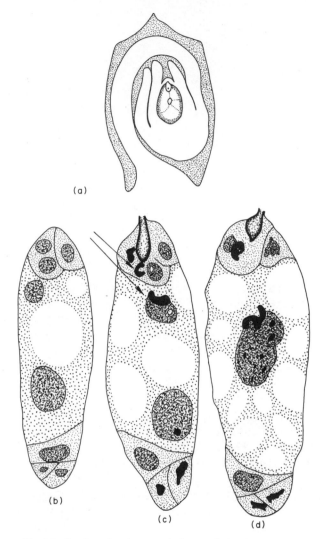

Fig. 1.1. Ovule and embryo sac before and after fertilization.
(a) General appearance of ovule of *Plumbago capensis*
(b), (c), (d) Embryo sac of *Lilium martagon* before and after fertilization
(b) Mature embryo sac
(c) Discharge of pollen tube into embryo sac (male nuclei), see arrows
(d) Contact of one male nucleus with egg nucleus and the other with two polar nuclei (after Maheshwari, 1950)

and, obviously, on the species (see Chapter 7). Other factors which determine the size and shape of seeds are the size of the embryo, the amount of endosperm present and to what extent other tissues participate in the seed structure. Some of the variety in seed form is illustrated in Fig. 1.2. Variability in seed shapes also exist within a given species and is then referred to as seed polymorphism. Characteristic of polymorphic seeds is that they differ not only in shape or colour, but also in their germination behaviour and dormancy. Among the better known examples are *Xanthium pennsyl-vanicum*, in which there are differences in germination requirements (Esashi and Leopold, 1968), *Salsola volkensii*, in which chlorophyllous and achlorophyllous

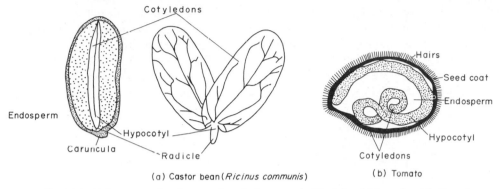

(a) Castor bean(*Ricinus communis*)

(b) Tomato

Fig. 1.3 (a, b) Structure of various seeds and of one-seeded fruit, in which seed and fruit coats are fused.

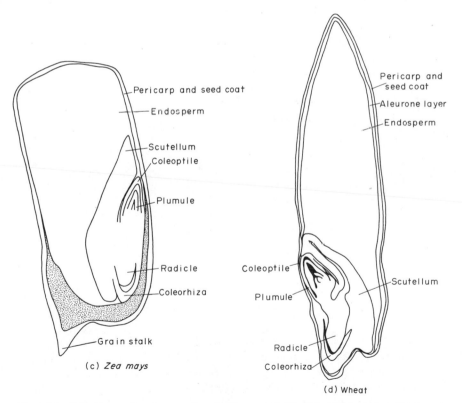

(c) *Zea mays*

(d) Wheat

Fig. 1.3 (c, d) Structure of various seeds and of one-seeded fruit, in which seed and fruit coats are fused.

embryos occur (Negbi and Tamari, 1964) and *Suaeda* in which the endosperm may be
either present or absent (Zohary, 1937, Roberts, 1972, Harper et al., 1970).

The normal seed contains materials which it utilizes during the process of
germination. These are frequently present in the endosperm, as in *Ricinus*, tomato,
maize and wheat (Fig. 1.3). The endosperm may contain a variety of storage material
such as starch, oils, proteins or hemicelluloses. But an endosperm is by no means
invariably present nor is it always the chief location of reserve materials. In many
plants the endosperm is greatly reduced, as in the Cruciferae, and in the orchids its
formation is entirely suppressed. In these cases, the reserve materials are present
elsewhere, for example in the cotyledons of the embryo, as in beans and lettuce (Fig.
1.3). In the orchids this is not the case. They contain virtually no reserve materials and

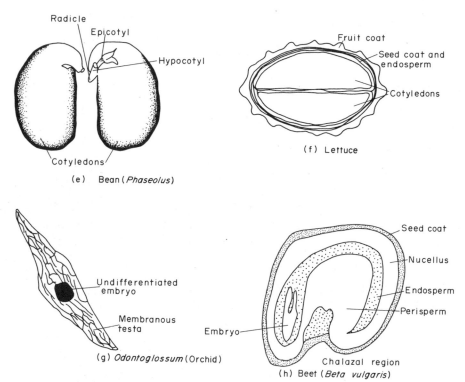

Fig. 1.3 (e, f, g, h) Structure of various seeds and of one-seeded fruit, in which seed and fruit
coats are fused.
(a), (e) after Troll, 1954; (b), (f), (h) after Hayward, 1938; (g) after McLean and Ivymey Cook,
1956.

therefore represent a very special case of seed structure. In some plants the storage
materials are contained in the perisperm, e.g. *Beta vulgaris* (Fig. 1.3). The perisperm
originates from the nucellus, e.g. in the Caryophyllaceae and in *Coffea*, and not from
the embryo sac as in the case for the endosperm. The structure of seeds as it appears
under the scanning electron microscope is shown in Fig. 1.3i, j.

Fig. 1.2. Illustration of forms of seeds (all scales in mm).

(a) *Xanthium strumarium*    (b) *Pisum sativum*    (c) *Ricinus communis*
(d) *Pancratium maritimum*    (e) *Citrullus colocynthis*    (f) *Phaseolus coccineus*
(g) *Raphanus sativus*    (h) *Phaseolus vulgaris*

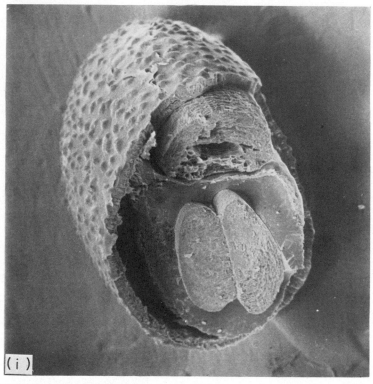

(i)

Fig. 1.3(i) Structure of seed of *Trigonella arabica* Del. Seed cut. Magnification ×67. (Courtesy Dr. Y. Gutterman). Scanning electron micrograph.

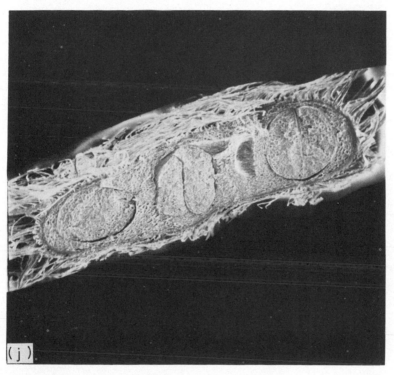

Fig. 1.3(j) Structure of seed of tomato. Seed cut. Magnification ×40. Note that cotyledons have been cut twice due to curled position of embryo. (Courtesy Mrs. I. Marbach). Scanning electron micrograph.

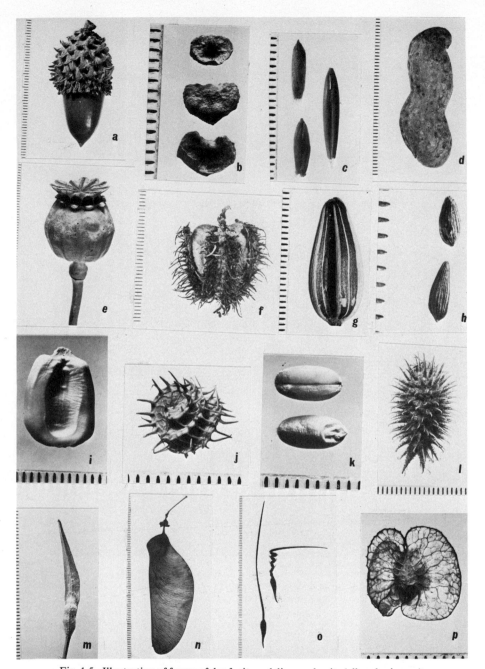

Fig. 1.5. Illustration of forms of dry fruits and dispersal units (all scales in mm)

(a) *Quercus ithaburensis*        (b) *Atriplex halimus*        (c) *Avena sativa*
(d) *Arachis hypogea*             (e) *Papaver orientalis*      (f) *Ricinus communis*
(g) *Helianthus annuus*           (h) *Lactuca sativa* var. Grand Rapids    (i) *Zea mays*
(j) *Medicago polymorpha*         (k) *Triticum vulgare*        (l) *Xanthium strumarium*
(m) *Sinapis alba*                (n) *Tipuana tipu*            (o) *Erodium maximum*
(p) *Rumex rosea*

Seeds are formed in the ovary and this develops into the fruit. Fruits arising primarily from the ovary are known as "true" fruits, while fruits in which other structures or several ovaries and their related structures participate, are often termed "false". Both types of fruits may be dry or fleshy. In many plants the integuments and the ovary wall are completely fused, so that the seed and fruit are in fact one entity as is the case of the grains of the grasses and in lettuce (Fig. 1.4). In other cases additional

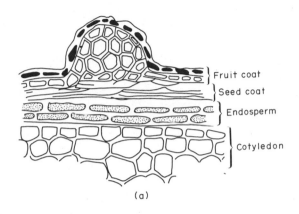

Fruit coat
Seed coat
Endosperm
Cotyledon

(a)

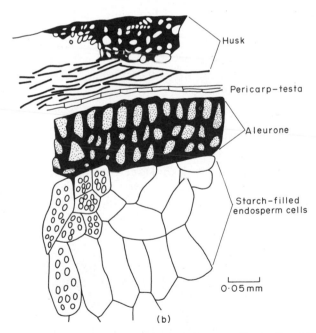

Husk

Pericarp–testa

Aleurone

Starch-filled endosperm cells

0·05 mm

(b)

Fig. 1.4. Fusion of seed and fruit coat in lettuce and barley "seeds".
(a) Lettuce (after Hayward, 1938).
(b) Barley (after McLeod, 1960).

parts of the plants, such as glumes, bracts and neighbouring sterile florets remain attached to the fruit and form a bigger dispersal unit. The exact nature and classification of these organs need not concern us here. A few of the very variable seed-bearing structures are, however, illustrated in Fig. 1.5.

The embryo consists of a radicle, a plumule or epicotyl, one or more cotyledons and a hypocotyl which connects the radicle and the plumule. The embryo may be variously located within the seed and may either fill the seed almost completely, as in the Rosaceae and Cruciferae, or it may be almost rudimentary, as in the Ranunculaceae. The classification of seeds is often based on the ratio of the size of the embryo to that of the storage tissue.

The process of germination leads eventually to the development of the embryo into a seedling. Seedlings are classified as *epigeal*, in which the cotyledons are above ground and are usually photosynthetic, and *hypogeal* in which the cotyledons remain below ground. In this latter case the cotyledons are either themselves the source of reserves for the seedling or receive materials from the endosperm. The size, shape and relative importance of the various tissues of seedlings are as variable as those of the seeds and fruits, and we will not attempt to classify them here. A few typical seedling forms are shown in Fig. 1.6. These indicate the large morphological variability which is met during germination.

In many plants special absorbing tissues develop which take up the reserve materials from the endosperm. In the onion the tip of the cotyledon is actually embedded in the endosperm and withdraws the materials from it (Fig. 1.6). The cotyledon with the seed attached is eventually raised above the ground and becomes green. In *Tradescantia* also the cotyledon is embedded in the endosperm, but the emerging organ is the coleoptile which is partly joined to the cotyledon and partly to the hypocotyl. In the Gramineae specialization has proceeded further and the scutellum fulfils the special function of a haustorium, which withdraws substances from the endosperm. In the various palms the haustorium, formed from the tip of the cotyledon, has become greatly enlarged. In a few cases the tip of the primary root rather than the scutellum of the cotyledon becomes the haustorium.

Such complications are absent in most of the dicotyledonous plants. Some straightforward examples are provided by the castor oil bean, *Ricinus communis*, which has epigeal germination with an endosperm, and *Phaseolus vulgaris* (French bean) which is epigeal without endospermic nutrition. In contrast, *Phaseolus multiflorus* is hypogeal and non-endospermic, while examples of hypogeal endospermic germination are provided by *Hevea* among the dicotyledons and *Zea mays* by the monocotyledons (Fig. 1.6).

After this brief discussion of forms of seeds and of germination we are now in a position to try and define germination and to begin a discussion of the processes which lead up to it. Germination of the seed of the higher plant we may regard as that consecutive number of steps which causes a quiescent seed, with a low water content, to show a rise in its general metabolic activity and to initiate the formation of a seedling from the embryo. The exact stage at which germination ends and growth begins is extremely difficult to define. This is particularly difficult because we identify germination by the protrusion of some part of the embryo from the seed coat, which in

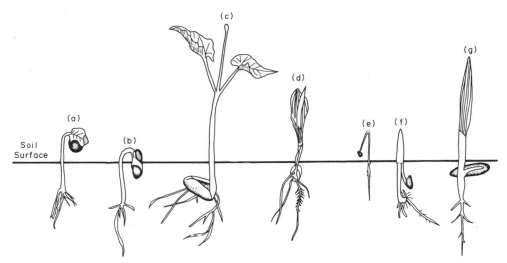

Fig. 1.6. Seedling morphology.

(a) *Ricinus communis* (epigeal)
(b) *Phaseolus vulgaris* (epigeal)
(c) *Phaseolus multiflorus* (hypogeal)
(d) *Zea mays* (hypogeal)
(e) Onion (epigeal)
(f) *Tradescantia virginica*
(g) *Phoenix dactylifera*

itself is already a result of growth. There is no general rule as to which part of the embryo first pierces the seed coat. In many seeds this is the radicle and therefore germination is frequently equated with root protrusion. However, in some seeds it is the shoot which protrudes first, for example in *Salsola*. The piercing of the seed coat by part of the embryo could be caused by cell division, cell elongation or both. In other words, the protrusion of part of the embryo through the seed coat is in fact the result of growth. Not all definitions would include cell division as such in the term growth. However, cell division is usually followed by an enlargement of the daughter cells and therefore constitutes growth. It is uncertain whether the process of germination can be equated with the processes of growth which lead to protrusion of part of the embryo through the seed coat. Probably the fundamental processes which cause germination are different from those of growth.

A further question which arises is whether cell division or cell elongation is the first process which occurs when the embryo pierces the seed coat. In some seeds it has been shown that cell division occurs first and is followed by cell elongation, while in other cases the reverse has been observed. Thus in lettuce seeds which are germinated at 26°C, cell division begins after some 12–14 hours and at about the same time cell elongation in root cells can also be observed (Evenari *et al.*, 1957).

In cherry seeds the changes in cell number and length of embryonic organs was followed during after-ripening under conditions of stratification at 5°C (Pollock and Olney, 1959). The length of the embryonic axis increased by some 7 per cent during stratification at 5°C. This increase in length was attributed to a combination of cell division and elongation. Changes at 5°C and 25°C were similar for the first 4 weeks of stratification but following this, axis length increased much more at 5°C than at 25°C. At the latter temperature no germination was obtained. Thus cherry and lettuce seem to be similar with regard to the more or less simultaneous occurrence of elongation and division. In germinating seeds of *Pinus thunbergii*, on the other hand, it seems that cell division precedes cell elongation, but the evidence is not entirely clear (Goo, 1952). Cell division occurs during a period when there is very little water uptake and before germination. In *Pinus lambertiana*, division and elongation were visually observed simultaneously. However thymidine incorporation occurred much earlier than cell elongation, indicating that cell division might precede cell elongation (Berlyn, 1972).

In contrast in the germination of *Zea mays*, the first change is cell enlargement in the coleorhiza. Cell division in the radicle occurs later, when the latter breaks through the coleorhiza and the seed coat. It is the coleorhiza which first pierces the seed coat (Toole, 1924). A similar situation apparently exists in barley.

A further attempt to differentiate between division and elongation during germination of lettuce has been made by Haber and Luippold (1960). Gamma radiation and low temperatures were used to delay cell division and hypertonic solutions of mannitol to arrest cell elongation. Despite the fact that irradiation and low temperatures prevented mitosis, germination as observed by root protrusion occurred. When, however, cell elongation was prevented by hypertonic solutions germination was prevented, although cell division could be clearly noted. Haber and Luippold conclude that division and cell elongation are affected by different factors and that root protrusion is primarily the result of cell elongation. Cell division only serves to increase the number of cells which can elongate. It is by no means clear whether these observations also apply to other seeds.

Cell division and cell elongation only occur once the cells of the seed have been activated in some way, so as to permit the control of these processes by various factors. Although it appears that for visible germination only one of these processes is necessary, there can be no doubt that for normal development and growth of the seedling both cell division and cell elongation are essential. It is thus possible to differentiate between these growth processes and the process of activation, which precedes growth and which may be termed germination.

However we will not, during the discussion of germination, attempt to make such a precise differentiation but rather will try to describe all those processes which take place up to seedling formation.

## Bibliography

Berlyn, G. P. (1972) in *Seed Biology*, vol. 1, p. 223. (ed. T. T. Kozlowski), Acad. Press, N.Y.
Bhatnagar, S. P. and Johri, B. M. (1972) in *Seed Biology*, vol. 1, p. 78. (ed. T. T. Kozlowski), Acad. Press, N.Y.
Esashi, Y. and Leopold, C. (1968) *Plant Physiol.* **43**, 871.
Evenari, M., Klein, S., Anchori, H. and Feinbrunn, N. (1957) *Bull. Res. Council, Israel,* **6D**, 33.
Goo, M. (1952) *J. of Jap. Forestry Soc.* **34**, 3.
Haber, A. H. and Luippold, H. J. (1960) *Plant Physiol.* **35**, 168.
Harper, J. L., Lovell, P. H. and Moore, K. G. (1970) *Ann. Rev. Ecol. Syst.* **1**, 327.
Hayward, H. E. (1938) *The Structure of Economic Plants.* MacMillan, New York.

McLean, R. M. and Ivimey-Cook, W. R. (1956) *Textbook of Theoretical Botany*, vol. **2**, Longmans Green, London.
McLeod, A. M. (1960) Wallerstein Lab. Comm. **23**, 87.
Maheshwari, P. (1950) *An Introduction to the Embryology of Angiosperms*, McGraw-Hill, New York.
Negbi, M. and Tamari, B. (1963) *Is. J. Bot.* **12**, 124.
Pollock, B. M. and Olney, H. O. (1959) *Plant Physiol.* **34**, 131.
Roberts, E. H. (1973) in *Seed viability*, p. 321. (ed. E. H. Roberts), Chapman and Hall, London.
Toole, E. H. (1924) *Amer. J. Bot.* **11**, 325.
Troll, W. (1954) *Praktische Einführung in die Pflanzenmorphologie*, Gustav Fisher, Jena.
Zohary, M. (1937) *Sond. Bei. Bot. Cent.* **56A**.

Chapter 2

# CHEMICAL COMPOSITION OF SEEDS

The chemical composition of seeds shows the same variability as is seen in other plant characteristics. The compounds found can be divided roughly into two groups, viz. (1) the normal constituents which are likely to occur in every plant tissue and (2) storage materials, which are frequently present in seeds in very large amounts. In addition, a large number of secondary plant products may be present in seeds. Most of the compounds which occur in them are not normally different from those found in the plant, but seed proteins differ in chemical composition and properties from those found in other plant tissues. The occurrence of large quantities of lipids differentiates seeds from all other plant tissues except certain fruits, since lipids do not usually occur in large amounts in other plant tissues.

Seeds can be divided into those whose main storage material is carbohydrate and those whose main storage material is lipid. Lipid-containing seeds are by far the bigger of these two groups, although among economically important seeds this preponderance does not occur to the same extent. Seeds containing proteins can belong to either group. Almost no seeds are known in which the predominant storage material is protein, although soybeans are an exception and *Machaerium acutifolium* has been reported to contain 66 per cent protein (Coutinho and Struffaldi, 1972). Precise information about seed composition is available chiefly for seeds which are used either for food or in industry. The chemical composition of seeds of various vegetable crops is less well known, while the information about the composition of the seeds of wild plants is extremely scant.

The constituents of seeds are determined genetically, but the relative amounts of these constituents are sometimes dependent on environmental factors such as mineral nutrition and climate. Thus Iwanoff (1927) showed that the protein content of wheat varied in various parts of Russia while that of peas was constant (Table 2.1). Modern plant breeding practice has permitted selection for quantitative differences in seed constituents. Thus soybeans have been bred for high protein content, flax for oil content and wheat both for protein and starch content. In some interesting breeding experiments, Woodworth *et al.* (1952) selected maize for various contents of protein and oil. Starting with maize containing 4·7 per cent oil and 10·9 per cent protein they obtained, after fifty generations of selection, four varieties having 15·4 or 1·0 per cent oil and 19·5 or 4·9 per cent protein. By selective breeding the composition of proteins can also be changed. Thus mutants of maize have been obtained in which the lysine and tryptophane content of the endosperm was almost doubled (Mertz *et al.*, 1966).

Tables 2.2, 2.3, 2.4 and 2.5 illustrate some of the variability in seed composition.

Table 2.1—Protein Content of Wheat (*Triticum vulgare*, var. *albidum*) and Peas (*Pisum sativum*, var. *vulgare*) Grown in Various Parts of Russia and Harvested in 1924
(After Iwanoff, 1927)

| Place | Longitude | Latitude | Protein content % dry weight | |
| --- | --- | --- | --- | --- |
| | | | Wheat | Peas |
| Severo Dvinsk | 46°18′ | 60°46′ | 11·92 | 27·00 |
| Moscow | 37°20′ | 55°48′ | 14·30 | 26.56 |
| Kiev | 30°28′ | 50°27′ | 19·32 | 30·37 |
| Saratov | 45°45′ | 51°37′ | 21·01 | 30·37 |
| Omsk | 70°32′ | 55°11′ | 18·69 | 28·12 |
| Krasnojarsk | 92°52′ | 56°01′ | 19·03 | 26·06 |
| Vladivistok | 131°57′ | 43°05′ | 11·86 | 26·87 |

Table 2.2—Chemical Composition of Seeds
(After Wehmer, 1929, Anderson and Kulp, 1921; Gutlin Schmitz, 1957; and Czapek, 1905)

| | Percentage of air-dry seeds | | | | |
| --- | --- | --- | --- | --- | --- |
| | Carbohydrates | | | | |
| | Starch | Sugar | Proteins | Fats | Lecithin |
| *Zea mays* | 50–70 | 1–4 | 10·0 | 5 | |
| *Pisum sativum* | 30–40 | 4–6 | 20·0 | 2 | |
| *Arachis hypogea* | 8–21 | 4–12 | 20–30·0 | 40–50 | 1·2 |
| *Helianthus annuus* | 0 | 2 | 25·0 | 45–50 | |
| *Ricinus communis* | 0 | 0 | 18·0 | 64 | |
| *Acer saccharinum* | 42 | 20 | 27·5 | 4 | |
| *Triticum* | 60–75 | | 13·3 | 2·0 | 0·7 |
| *Fagopyrum esculentum* | | 72·0 | 10·0 | 2·0 | 0·5 |
| *Chenopodium quinoa* | | 48·0 | 19·0 | 5·0 | |
| *Aesculus hippocastanum* | | 68·0 | 7·0 | 5·0 | |
| *Castanea vesca* | | 42·0 | 4·0 | 3·0 | |
| *Quercus pendunculata* | | 47·0 | 3·0 | 3·0 | |
| *Linum usitatissimum* | | 23·0 | 23·0 | 34·0 | |
| *Brassica rapa* | | 25·0 | 20·0 | 34·0 | |
| *Papaver somniferum* | | 19·0 | 20·0 | 41·0 | |
| *Cannabis sativa* | | 21·0 | 18·0 | 33·0 | 0·4 |
| *Amygdalus communis* | | 8·0 | 24·0 | 53·0 | 0·9 |
| *Aleurites moluccana* | | 5·0 | 21·0 | 62 | |

An example of the composition of a wild plant, *Amaranthus*, is shown in Table 2.6. From these tables it will be seen that in addition to carbohydrates, proteins and lipids seeds contain minerals, tannins, phosphorous compounds, etc. A classification of seed constituents is difficult because of the large variety of compounds met with, but a rough breakdown into carbohydrates, proteins, lipids and other compounds seems permissible.

Table 2.3—Chemical Composition of Lettuce Seeds
(Original)

|  | air-dry seeds (mg/g) |
|---|---|
| Total dry weight | 960·0 |
| Ash | 46·0 |
| Phytic Acid | 20·0 |
| Sucrose | 30·0 |
| Glucose | 2·0 |
| Fat | 370·0 |
| Total nitrogen | 40·0 |
| Protein nitrogen | 37·0 |
| Soluble nitrogen | 1·0 |
| Riboflavin | 0·012 |
| Ascorbic acid | 0·29 |
| Carotene | 0·004 |
| Total P (free and bound) | 8·5–14 |

Table 2.4—Chemical Composition of Soybeans
(Compiled from data of Morse, 1950)

| A. Major Constituents | % of moisture-free basis |
|---|---|
| Moisture | 8·0 |
| Ash | 4·6 |
| Fat | 18·0 |
| Fibre | 3·5 |
| Protein | 40·0 |
| Pentosans | 4·4 |
| Sugars | 7·0 |
| Starch-like substances | 5·6 |
| Phosphorous | 0·63 |
| Potassium | 1·67 |
| Calcium | 0·26 |

| B. Mineral Constituents | % air-dry weight of seeds |
|---|---|
| Magnesium | 0·22 |
| Sulphur | 0·41 |
| Chlorine | 0·024 |
| Iodine | trace |
| Sodium | 0·34 |
| Manganese | 0·0028 |
| Zinc | 0·0022 |
| Aluminium | 0·0007 |
| Copper | 0·0012 |
| Iron | 0·0097 |

| C. Vitamins | $\mu$g/g |
|---|---|
| Thiamine | 17·5 |
| Riboflavine | 3·6 |
| Pyridoxine | 11·8 |
| Nicotinic Acid | 22·4 |
| Pantothenic Acid | 21·5 |
| Inositol | 2291·0 |
| Biotin | 0·8 |

Table 2.5—Composition of *Vicia faba* Seeds
(From data of White, 1966)

| | | | |
|---|---|---|---|
| Total Dry matter % Fresh weight | 90·03% | | |
| Dry matter in Embryo + Cotyledons | 86·28% | | |
| Composition of Embryo + Cotyledons | | | |
| Carbohydrates | 56·67% | | |
| Total Nitrogen | 35·8 | (Insoluble N₂ | 31·69%) |
| Fat | 2·03 | | |
| Organic Acid | 1% | | |
| Ash | 9·29 | | |

| | |
|---|---|
| Composition of carbohydrate fraction | |
| Hexose | 0·16% |
| Sucrose | 4·02% |
| Starch | 42·48% |
| Pectin | 1·69% |
| Hemicellulose | 6·66% |
| Cellulose | 1·66% |

Table 2.6—Composition of Seeds of *Amaranthus retroflexus*
(Woo, 1919)

| | % air-dry seeds |
|---|---|
| Water | 8·6 |
| Lipids | 7·8 |
| Polysaccharides | 47·2 |
| Reducing sugars | none |
| Non-reducing sugars (after hydrolysis) | 1·2 |
| Nitrogen | 2·5 |
| Protein | 15·0 (Soluble 3·0 Insoluble 12·0) |
| Ash | 4·2 |

## I. Carbohydrates

The two chief storage carbohydrates are starch and hemicelluloses. The former is found in all the food grains and in legumes, while hemicelluloses, both pentosan and hexosan, occur in the endosperm of the palms as well as in the cotyledons of lupins, *Primula* and *Impatiens*. In addition, however, many other carbohydrates occur in seeds, not necessarily as storage materials. Thus in many seeds various polyuronides occur frequently in the seed coat. These mucilages are possibly connected with seed dispersal and water uptake during germination (Young and Evans, 1973).

Mucilages occur either on the seed surface, as in flax, or in special cells in the seed coat, as in *Brassica alba*. *Plantago* seeds constitute a commercial source for mucilages. Chemically the mucilages are polyuronides, mainly galacturonides, or galactomannans or complex compounds containing both galactomannans and uronic acids (Tookey *et al.*, 1962). In addition, various sugars, hexoses and pentoses, have been isolated from mucilages. The polyuronides are frequently associated with proteins. Occasionally cellulose fibres are found in mucilages. Galactomannans also are present as the main reserve carbohydrate especially in the endosperm of Leguminoseae, for example *Trigonella* (Reid and Meier, 1970). In some species of the Liliaceae and Iridaceae, galactoglucomannans are present and even more complex carbohydrates have been reported which contain galactose, mannose, glucose and xylose.

Pectins are a normal constituent of plant cells and consequently also of seeds. Mono-, di-, tri- and oligo-saccharides, e.g. glucose, fructose, sucrose, raffinose and stachyose, occur in greater or smaller amounts in most seeds. In barley a considerable amount of fructosans occurs. The carbohydrate composition of barley seeds is shown in Table 2.7, as is the division of these constituents among various parts of the

Table 2.7—Carbohydrate Composition of Barley
(McLeod, 1960)

| | % dry-weight of tissue | | |
| | Husk | Embryo | Endosperm |
| --- | --- | --- | --- |
| Sucrose | 0 | 14·0 | 0·2 |
| Raffinose | 0 | 10·0 | 0·08 |
| Hexoses | 0 | 0·2 | 0·26 |
| Total glucan | <0·02 | 0 | 1·7 |
| Total pentosan | 3·6 | 0·4 | 1·0 |
| Galactan | 0 | 0·3 | 0 |
| Uronic acid | + | + | 0 |
| Crude cellulose | 30 | 7·0 | 0·4 |

seed. Ivory nuts contain mannans, and galactans occur in *Lupinus albus* and *Strychnos*. Many secondary plant products are present in seeds in the form of glycosides. Thus bitter almonds contain amygdalin, mandelo-nitrile gentiobioside. Black mustard seeds contain sinigrin, the glycoside of a mustard oil, and *Nigella* contains damasenine. Various alkaloids, tannins and leucoanthocyanins also occur as glycosides in seeds.

## II. Lipids

Lipids are generally present in the form of the glycerides of fatty acids:

$$CH_2OR_1$$
$$|$$
$$CHOR_2$$
$$|$$
$$CH_2OR_3$$

where $R_1$, $R_2$ and $R_3$ may be the same, or different fatty acids. Most seed fatty acids are unsaturated and the most commonly-occurring ones are oleic, linoleic and linolenic acids. In addition, however, other organic acids, both saturated and unsaturated, such as acetic, butyric, palmitic, stearic, lauric and myristic acids and many others, occur as glycerides. Ground-nuts contain the glyceride of arachidic acid. The lipids are found both as fats and as oils, depending on the relative amounts of saturated and unsaturated fatty acids occurring in the glycerides.

In recent years improved analytical methods have shown the presence of very diverse seed triglycerides in seed oils. In addition to the well known mono- and di-unsaturated acids, polyolefinic acids and acytelenic acids are now known to occur in seed. Oxygenated fatty acids seem to be present quite frequently (Wolff, 1966).

Other lipid materials found in seeds are esters of higher alcohols, sterols, phospholipids and glycolipids, tocopherols and squalenc.

An example of the composition of seed lipids from soybeans is given in Table 2.8.

Table 2.8—Composition of Lipids from Soybeans
(Compiled from data of Daubert, 1950)

|  | Total | % dry weight of bean |
|---|---|---|
| Saturated Fatty Acids | 11–13 | |
| Myristic | | 0·1–0·4 |
| Palmitic | | 6·5–9·8 |
| Stearic | | 2·4–5·5 |
| Arachidic | | 0·2–0·9 |
| Unsaturated Fatty Acids | 88 | |
| Oleic | | 11–60 |
| Linoleic | | 25–63 |
| Tetradecenoic | | trace |
| Hexadecenoic | | trace |
| Average Iodine Value | 102–150 | |
| Other Constituents | | |
| Carotene (chiefly $\beta$) | | |
| Chlorophyll | | |
| Sterols | | |
| Phytosterolins | | |
| Tocopherols | | |
| Phosphatides | | |

A highly toxic oil has been isolated from seeds of *Dichapetalum toxicarium*. This lipid has been shown to contain fluoro-oleic acid (Peters *et al.*, 1960). The compound is so toxic that the seeds of this plant are often referred to as "ratsbane". It is interesting that the leaves of this plant contain fluoro-acetic acid.

## III. Proteins

One of the characteristics of proteins of seeds is that while some of them are metabolically active, such as the enzyme proteins, many of them are metabolically inactive. These latter are the storage proteins which vary according to the species. Thus, in wheat, at least four different proteins occur, glutelins, prolamins, globulins and albumins. The glutelins and prolamins form the major component of the protein, while the globulins contribute only 6–10 per cent of the total and the albumins 3–5 per cent. The total active proteins, globulins and albumins account for no more than 15 per cent of the total proteins in wheat. The distribution between metabolically-active and inactive proteins is similar in most cereals.

The division of the storage proteins into glutelins and prolamins is somewhat arbitrary and is based primarily on the differential solubility of the proteins in weak acids and alkalis. It is therefore preferable to refer to definite proteins whose properties have been investigated. The best-known prolamins are those found in wheat-gliadin, in barley-hordein and in maize-zein. Glutelin from wheat is probably a mixture of a number of proteins. Other glutelin proteins, such as zecanin, hordenine and oryzenin have not been studied by modern methods and it is therefore not certain whether single proteins or mixtures of proteins are involved. In dicotyledonous plants prolamins seem to be almost absent. Glutelins are sometimes absent and sometimes constitute up to 50 per cent of the total proteins in dicotyledonous seeds. Albumins and globulins in these seeds are usually well-defined and a number of the globulins have been obtained in crystalline form after extraction with hot salt solutions. Among those which have been investigated in detail are legumin and vicillin from peas, which have been obtained electrophoretically homogenous, arachin and conarachin from peanuts, glycinin from soybeans and edestin from hemp seeds. Legumin and vicillin from *Phaseolus aureus* are glycoproteins, containing neutral sugars and a small amount of glucosamine. Both proteins are made up of non-identical subunits, three for legumin and four for vicillin (Ericson and Chrispeels, 1973). The storage proteins usually occur in well defined organelles, which can be distinguished in the electron microscope. Generally these are referred to as protein bodies or microbodies.

Seed storage-proteins generally have a high nitrogen content, high proline content and are often low in their content of lysine, tryptophan and methionine. Table 2.9

Table 2.9—The Protein Content of Various Seeds
(From Brohult and Sandegren, 1954)

| | Total protein | Various fractions as % of total protein | | | |
|---|---|---|---|---|---|
| | % dry seeds | Albumin | Globulin | Prolamin | Glutelin |
| *Triticum vulgare* | 10–15 | 3–5 | 6–10 | 40–50 | 30–40 |
| *Hordeum vulgare* | 10–16 | 3–4 | 10–20 | 35–45 | 35–45 |
| *Avena sativa* | 8–14 | 5–10 | 80 | 10–15 | 5 |
| *Secale cereale* | 9–14 | 5–10 | 5–10 | 30–50 | 30–50 |
| *Cucurbita pepo* | 12 | very little | 92 | very little | small amounts |
| *Nicotiana sp.* | 33 | 24 | 26 | very little | 50 |
| *Gossypium herbaceum* | 20 | very little | 90 | very little | 10 |
| *Glycine hispida* | 30–50 | small amounts | 85–45 | very little | very little |
| *Lupinus luteus* | 40 | 1 | 78 | very little | 16 |

Table 2.10—Amino Acid Composition of Some Seed Proteins (Amino acid as % of
protein)
(From Brohult and Sandegren, 1954; and Circle, 1950)

|  | Gliadin | Zein | Legumin | Glycinin | Arachin |
|---|---|---|---|---|---|
| Total nitrogen | 17·0 | 16·2 | 17·9 | 17·0 | 18·3 |
| Glycine | 1·0 | 0 |  | 1·0 | 3·7 |
| Alanine | 2·0 | 11·5 |  |  | 5·0 |
| Serine | 4·8 | 7·8 |  |  | 5·3 |
| Threonine | 2·1 | 3·0 |  |  | 2·6 |
| Valine | 2·6 | 3·0 |  | 1·0 | 4·5 |
| Leucine | 6·7 | 24·0 |  | 8·5 | 7·8 |
| Isoleucine | (5·1) | 7·4 |  | 2·5 | 7·6 |
| Methionine | 1·6 | 2·3 |  | 1·8 | 0·7 |
| Cysteine and/or cystine | 2·5 | 1·0 |  | 1·1 | 1·5 |
| Proline | 13·2 | 10·5 |  | 4·0 | 7·0 |
| Phenylalanine | 6·3 | 6·5 |  | 4·0 | 6·3 |
| Tryptophan | 0 | 0 | 1·3 | 1·6 | 1·0 |
| Tyrosine | 3·3 | 5·3 | 4·3 | 1·8 | 5·5 |
| Histidine | 2·3 | 1·7 | 3·0 | 1·4 | 2·2 |
| Arginine | 2·7 | 1·8 | 13·1 | 8·1 | 14·0 |
| Lysine | 1·2 | 0 | 3·5 | 4·0 | 4·6 |
| Aspartic acid | 3·6 | 5·7 | 16·3 | 5·0 | (16·0) |
| Glutamic acid | 4·7 | 27·0 | 30·0 | 19·0 | 23·8 |

shows the protein content of some mono- and dicotyledonous seeds, and Table 2.10
shows the amino acid composition of some seed proteins.

Proteinase inhibitors, such as the soybean trypsin inhibitors, are of widespread
occurrence in seeds. These inhibitors are polypeptides or low molecular weight
proteins (Vogel *et al.*, 1968).

## IV. Other Components

In addition to the major compounds mentioned, seeds contain a large number of
other substances. All seeds contain a certain amount of minerals (see Table 2.4). The
mineral composition of seeds is essentially similar to that of the plant and usually
comprises all the essential and minor elements. The phosphorous composition of
*Amaranthus* is shown in Table 2.11. A special feature of many seeds is that a very
large part of the phosphate occurs as phytin, the calcium and magnesium salt of
inositol hexaphosphate. In lettuce seeds 50 per cent of the total phosphorus occurs
as phytin, 6–10 per cent as free phosphate and the remainder in other phosphorus-
containing compounds such as nucleotides, sugar-phosphates (20–25 per cent),
phospholipids, nucleoproteins and other compounds (20–25 per cent). In addition,
the nucleic acids constitute an extremely important part of the phosphorus-
containing compounds. The nucleic acids occur partly in their free form and partly in
the form of nucleoproteins. The ratio of RNA to DNA (ribonucleic acid to
deoxyribonucleic acid) in many seeds is approximately 10:1.

Table 2.11—Phosphorus Compounds in *Amaranthus retroflexus*
(Woo, 1919)

|  | P as % of total ash |
| --- | --- |
| Total P | 4·6 |
| Inorganic P | 0·13 |
| Lipid P | 0·18 |
| Soluble organic P | 0·33 |
| Phosphoprotein P | 1·8 |
| Nucleoprotein P | 2·5 |

The nitrogen content of seeds usually comprises, in addition to proteins, a certain amount of free amino acids and amides. The amides found are glutamine and asparagine, as well as γ-methylene glutamine, as for example in ground-nuts. The free amino acids found in seeds are usually the same as those forming part of the protein structure. Many of the non-protein amino acids, present in various tissues of plants, also occur in the seed of the plants, while others have been shown to occur especially in the seeds. Among these compounds are γ-methylene glutamic acid, γ-aminobutyric acid, β-pyrazol-l-ylalanine, lathyrine, pipecolic acid and many others. Some of these acids are restricted to a single species, while others are of very widespread distribution among many species (Bell, 1966, Fowden, 1970). Some of the non-protein amino acids are highly toxic to animals or man. There are some reports that the presence of at least some of these substances makes the seeds less edible to predators (Rehr *et al.*, 1973).

Other nitrogenous constituents of seeds are various alkaloids. Some examples are provided by piperine in *Piper nigrum* seeds, ricinine in castor oil beans, hyoscine in seeds of *Datura* and lupinidine (sparteine) in lupin seeds. It is interesting to note that the occurrence of some alkaloid in the plant does not necessarily indicate its presence in the seed in comparable amounts. A special case is that of *Coffea*, where the caffeine content of the bean is much higher than that of the plant. Cacao (*Theobroma cacao*) has a very high content of theobromine in the seeds as well as a smaller amount of caffeine. Strychnine and brucine are obtained from the seeds of *Strychnos nux-vomica* where they amount to about 2–3 per cent of the seeds.

Various organic acids such as tricarboxylic acid cycle intermediates, as well as malonic acid, have been detected in seeds of many species.

Phytosterols occur in a number of seeds. The best-known ones are the sitosterols and stigmasterols from soybeans. The latter is important pharmaceutically as it is used as a precursor of progesterone.

Although seeds contain a number of pigments, chlorophyll is usually absent although it occurs in the seeds of gymnosperms. Protochlorophyll, however, occurs in the Cucurbitaceae. Breakdown products of chlorophyll seem to be present in a number of seeds. Other pigments found are carotene and various other carotenoids. The seeds coats of many seeds contain anthocyanins or leucoanthocyanins. Flavonoid pigments are also known to be present in various seeds. Many of these compounds occur as the corresponding glycosides. Thus the seed coat of *Phaseolus vulgaris* may contain leuco-delphinidin, leuco-pelargonidin as well as delphinidin, petunidin and malvidin as the corresponding glycosides, and also at least two flavonol glycosides. Cotton seeds contain a yellow pigment, gossypol, in special pigment glands.

Various phenolic compounds such as coumarin derivatives, chlorogenic acid and simple phenols such as ferulic, caffeic and sinapic acids, occur in many seeds. These compounds may give rise, by oxidation and condensation, to pigments of the melanin type. Another phenolic type constituent is the tannins, e.g. in *Arachis* seeds.

Although the function of vitamins in seeds is as yet unknown, in all cases where seeds have been examined, vitamins have in fact been found to be present. The presence of vitamins is especially important in those seeds which are used for food or fodder. In these, detailed information on the vitamin content and especially the B component is available, and attempts are continuously made to raise the vitamin content of such seeds by selective breeding. Some figures are given in Table 2.12.

Table 2.12—The Vitamin Content of Some Seeds. The Figures are for the Air-dry Seeds (From Food Composition Tables, FAO, 1954)

| | mg/100 g | | | | I.U |
|---|---|---|---|---|---|
| | Thiamin | Riboflavin | Nicotinic Acid | Ascorbic Acid | Vit. A |
| Wheat (Durrum) | 0·45 | 0·13 | 5·4 | 0 | 0 |
| Rice (hulls removed) | 0·33 | 0·05 | 4·6 | 0 | 0 |
| Barley | 0·46 | 0·12 | 5·5 | 0 | 0 |
| Maize | 0·45 | 0·11 | 2·0 | 0 | 450 |
| Ground-nuts (shelled) | 0·84 | 0·12 | 16·0 | 0 | 30 |
| Soybean | 1·03 | 0·30 | 2·1 | 0 | 140 |
| Broad bean | 0·54 | 0·29 | 2·3 | 4 | 100 |
| Peas | 0·72 | 0·15 | 2·4 | 4 | 100 |
| Sunflower | 0·12 | 0·10 | 1·4 | 0 | 30 |
| Chestnuts (fresh) | 0·21 | 0·17 | 0·4 | 24 | 0 |

Tocopherols are present in the oil of many seeds. Most of the known vitamins have in fact been shown to occur in some kinds of seeds.

With the development of modern analytical methods and paper chromatographic techniques, linked with biological assays, it has become possible to study the growth substance content of various seeds. The occurrence of indolylacetic acid in a number of seeds has been demonstrated (Elkinowy and Raa, 1973), while in other cases various indole derivatives have been found. Gibberellic acid and gibberellin-like substances have been found in runner beans, in lettuce seeds, barley and in the seeds of many other plants (Moundes and Michael, 1973). Cytokinins, both free and bound occur in seeds. The first natural cytokinin, zeatin, was isolated from immature maize kernels. The plant growth inhibitor abscisic acid (ABA), has been found in seeds of *Fraxinus* and the achenes of *Rosa*. Probably it occurs in many species, particularly in those showing dormancy. Ethylene is formed in seeds very rapidly after the beginning of imbibition and must be added to the list of growth substance occurring in seeds. In addition, growth-promoting and growth-inhibiting substances have been shown to be present in various seed extracts, but their nature has not yet been elucidated. The possible importance of these substances in the regulation of dormancy will be discussed in a later chapter.

## Bibliography

Anderson, R. J. and Kulp, W. L. (1921) *N.Y. Agr. Expr. Sta. Bull.* **81**.

Bell, E. A. (1966) In *Comparative Biochemistry*, p. 195 (ed. T. Swain), Acad. Press, London.

Brohult, S. and Sandegren, E. (1954) In *The Proteins*, vol. **2A**, p. 487, Acad. Press, New York.

Circle, S. J. (1950) In *Soybeans and Soybean Products*, vol. **1**, p. 275, Interscience, New York.

Coutinho, L. M. and Struffaldi, Y. (1972) *Phyton* **29**, 25.

Czapek, F. (1905) *Die Biochemie der Pflanzen*, Gustav Fisher, Jena.

Daubert, B. F. (1950) In *Soybeans and Soybean Products*, vol. **1**, p. 157, Interscience, New York.

Elkinowy, M. and Raa, J. (1973) *Physiol. Plant.* **29**, 250.

Ericson, M. C. and Chrispeels, M. J. (1973) *Plant Physiol.* **52**, 98.

FAO Food Composition Tables, Minerals and Vitamins, Rome, Italy (1954).

Fowden, L. (1970) In *Progress in Phytochemistry*, vol. **2**, p. 203, Interscience, London.

Gutlin Schmitz, P. H. (1957) Dissertation of the University of Basel; Über die Änderung des Aneuring-ehaltes während der Keimung in Samen verschiedener Reserve Stoffe.

Iwanoff, N. N. (1927) *Biochem. Z.* **182**, 88.

Iwanoff, N. N. (1927) *Bull. Bot. and Plant Breeding*, **17**, 225.

McLeod, A. M. (1960) *Wallerstein Lab. Comm.* **23**, 87.

Mertz, E. T., Nelson, O. E., Bates, L. S. and Veron, O. A. (1966) *Adv. Chem.* no. **57**, p. 228. Amer. Chem. Soc.

Moundes, M. A. Kh. and Michael, G. (1973) *Physiol. Plant.* **29**, 274.

Morse, W. J. (1950) In *Soybeans and Soybean Products*, vol. **1**, p. 135, Interscience, New York.

Peters, R. A., Hall, R. J., Ward, P. F. and Sheppard, N. (1960) *Biochem. J.* **77**, 17.

Rehr, S. S., Bell, E. A., Janzen, D. H. and Feeny, P. P. (1973) *Biochem. Syst.* **1**, 63.

Reid, J. S. G. and Meier, H. (1970) *Phytochem.* **9**, 513.

Tookey, H. L., Lohmar, R. L., Wolff, I. A. and Jones, Q. (1962) *Agr. Found. Chem.* **10**, 131.

Vogel, R., Trautschold, J. and Werle, E. (1968) *Proteinase Inhibitors*, Acad. Press, New York.

Wehmer, C. (1929) *Die Pflanzenstoffe*, 2nd Edition, Jena.

White, H. L. (1966) *J. Expt. Bot.* **17**, 195.

Wolff, I. A. (1966) *Science* **154**, 1140.

Woo, M. L. (1919) *Bot. Gaz.* **68**, 313.

Woodworth, C. M., Leng, E. R. and Jugenheimer, R. W. (1952) *Agron. J.* **44**, 60.

Young, J. A. and Evans, R. A. (1973) *Weed Science* **21**, 52.

Chapter 3

# FACTORS AFFECTING GERMINATION

## I. Viability and Life Span of Seeds

Seeds are fairly resistant to extreme external conditions, provided they are in a state of desiccation. As a result seeds can retain their ability to germinate, or viability, for considerable periods. The length of time for which seeds can remain viable is extremely variable and depends both on the storage conditions and on the type of seed. In general, viability is retained best under conditions in which the metabolic activity of seeds is greatly reduced, i.e. low temperature and high carbon dioxide concentration. In addition, however, other factors are of great importance, particularly those which determine seed dormancy. The period for which seeds remain viable is determined genetically and by environmental factors. The latter will in fact have a decisive effect on the life span of any given seed, i.e. whether the seed will remain viable for the longest genetically possible period or whether it will lose its viability at some earlier stage.

Becquerel (1932 and 1934) tried to germinate seeds which were taken from the Herbaria of the National Museum in Paris, in order to estimate their life span. He came to the conclusion that *Mimosa glomerata* seeds remained viable for some 221 years, while various other leguminous seeds remained viable for periods of 100–150 years, e.g. *Astragalus massiliensis*, *Dioclea paucifera* and *Cassia bicapsularis*.

Turner (1933) tested seeds from collections kept at Kew. He also cites many examples in which life span was estimated from data on length of burial period and similar circumstantial information. Some of the data of Turner and Becquerel are given in Table 3.1. Many seeds (e.g. charlock) kept better, buried in soil, than in jars on the laboratory shelves.

In contrast to the seeds listed in Table 3.1, which have a very long life span, other seeds are characterized by a very short life span. For example *Acer saccharinum*, *Zigana aquatica*, *Salix japonica* and *S. pierotti* lose their viability within a week if

Table 3.1—Life Span of Seeds In Years
(From data of Becquerel 1932 and 1934; and Turner, 1933)

| | | | |
|---|---|---|---|
| *Anagallis foemina* | 60 | *Lotus uliginosus* | 81 |
| *Anthyllis vulneraria* | 90 | *Medicago orbicularis* | 78 |
| *Cassia bicapsularis* | 87 | *Nelumbium luteum* | 56 |
| *Cytisus biflorus* | 84 | *Stachys nepetifolia* | 77 |
| *Ipomoea sp.* | 43 | *Trifolium arvense* | 68 |
| *Lavatera olbia* | 64 | *Trifolium pratense* | 81 |
| *Lens esculenta* | 65 | *Trifolium striatum* | 90 |

kept in air. *Ulmus campestris* and *U. americana* remain viable for about 6 months. *Hevea*, sugar cane and *Boehea*, *Thea*, cocos and other tropical crop seeds, remain viable for less than a year (Crocker, 1938).

Several attempts have been made to establish the viability of seeds by controlled experiments. The seeds were usually buried in the soil in some suitable containers and samples removed at different periods. Thus Beal began experiments in 1879 and these were continued for a period of 30 years, seeds being removed and tested every 5 years. After 30 years a considerable number of species was still viable, namely *Amaranthus retroflexus*, *Brassica nigra*, *Capsella bursa-pastoris*, *Lepidium virginicum*, *Oenothera biennis*, *Rumex crispus* and *Setaria media*. Subsequently, seeds were tested every 10 years. After 90 years of burial, seeds of *Verbascum blattaria* were still viable, germinated and the plants produced normal seeds (Kivilaan and Bandurski, 1973). Similar experiments were conducted by Duvel (1905) and summarized by Goss in 1924. The final results of this experiment were reported by Toole and Brown (1946). A long-term experiment plan to last more than 300 years has been initiated by Went and Munz (1949).

Both in the experiments of Beal and those of Becquerel the seeds were maintained under relatively dry conditions during storage. It is known today that seeds generally remain viable for longer periods if they are dry. For example, lettuce seeds kept much longer if their moisture was reduced from air-dry to half this value (in Ithaca U.S.A.) (Griffiths, 1942). Moisture content was more critical than temperature during storage. Raising moisture content from 5 to 10 per cent caused a more rapid loss of viability than a temperature rise from 20° to 40°C. Similar data have been obtained for clover. Here the seeds remained viable for 3 years at all temperatures up to 38°C when their moisture content was 6 per cent. At 8 per cent moisture they remained viable except at this highest temperature. When the moisture content of the seeds was raised to 12–16 per cent the seeds remained viable only at 30°C. At 16 per cent moisture content viability at 22°C fell within 3 months (Ching *et al.*, 1959). The moisture content of seeds is determined, at least in part, by the relative humidity of the air and by the temperature of storage. Generally as the temperature and relative humidity increase moisture content of the seeds increases up to a maximal value and then falls again as the temperature increases further. Loss of germinability usually occurs before the maximal water content is reached (Barton, 1951).

An interesting attempt has recently been made to express more exactly the relationship between viability, storage temperature and moisture content of cereal seeds (Roberts, 1972). Roberts was able to express all the known data on viability by a simple mathematical relationship:

$$\log \bar{p} = K_v - C_1 m - C_2 t$$

where $\bar{p}$ is the mean viability period of the seeds, $t$ is the temperature in °C, $m$ the moisture as per cent and $K$ and $C$ are constants. These constants were calculated from experiments with wheat, but they appeared to fit also data obtained from oats and barley. From this expression it is possible to predict the expected viability and life span of a given cereal seed under almost all storage conditions. Nomographs have been made to predict the behaviour of a number of species (Roberts, 1972).

Although from the above data it appears that dry conditions are essential for retention of viability, many seeds remain viable when submerged in water. Shull (1914) was able to show that 11 out of 58 species tested were still viable after $4\frac{1}{2}$ years' submergence. Some of the results obtained by Shull are summarized in Table 3.2 which lists the species of seeds submerged under water and those subsequently found to germinate.

Table 3.2—Viability of Various Seeds When Kept Submerged Under Water for Various Periods of Time (+ seeds germinated after indicated period of submergence) (Shull, 1914)

| | 130 days | 18 months | 30 months | 54 months | 7 years |
|---|---|---|---|---|---|
| *Agrimonia hirsuta* | | | | | |
| *Asclepias syriaca* | | | | + | |
| *Chenopodium album* | | | + | | |
| *Circaea lutetiana* | | | | | |
| *Geum carolinianum* | | | | | |
| *Hieracium sp.* | | | | | |
| *Juncus bufonius* | + | | | | + |
| *Juncus tenuis* | + | + | | + | + |
| *Lappa minor* | | | + | | |
| *Muhlenbergia diffusa* | | + | + | + | |
| *Phryma leptostachya* | | | | | |
| *Plantago rugelii* | + | + | + | + | |
| *Polygonum arifolium* | | | | | |
| *Polygonum virginianum* | | | | | |
| *Rhus glabra* | | | | | |
| *Sanicula marylandica* | | | | | |
| *Sium cicutaefolium* | | + | + | + | + |
| *Solidago rugosa* | | | | + | |
| *Sparganium androcladum* | | | | + | |
| *Unifolium canadense* | | | | | |
| *Verbena urticaefolia* | | + | + | + | |
| *Washingtonia longistyllis* | | | | | |

According to Villiers (1974) lettuce seeds stored fully imbibed at 30° lost their viability less rapidly than seeds with a moisture content of 5·1 per cent. As the moisture content rose to 13·5 per cent viability was lost after 2 months. Similar results were obtained for *Fraxinus americana*. He ascribes these results to the operation of repair mechanisms in the fully imbibed seeds which cannot operate in partially hydrated ones.

From the foregoing it can be seen that the storage conditions required to maintain viability for different seeds are different. Thus cases are known where drying causes very rapid loss in viability, e.g. *Acer saccharinum*, while in other cases only on drying will the seeds remain viable. Similar differences in storage conditions are known for oxygen and carbon dioxide concentration. Therefore it is very difficult to formulate any general rule for favourable storage conditions. This problem is also discussed by Owen (1956) and by Roberts (1972).

Even under favourable storage conditions many seeds are relatively short-lived, for example the seeds of many trees and of various vegetables. Such a loss of viability is not a sudden abrupt failure to germinate of all the seeds in a certain population. Rather the percentage of seeds which will germinate in any given population will slowly decrease. Moreover, even if a seed loses its viability this does not imply that all metabolic processes stop together or that all enzymes are inactivated. Only the sum total of processes which lead to germination no longer operates. This was illustrated for cocoa beans by Holden (Table 3.3). For this reason

Table 3.3—Enzyme Activity and Viability in Cocoa Beans
The beans were allowed to ferment which raises their
temperature and eventually kills them
(After Holden, 1959)

|  | Enzyme activity as % of unfermented beans | | | |
|---|---|---|---|---|
| Time of fermentation | 20 | 44 | 68 | 92 hr |
| Temp. °C | 33 | 42 | 44 | 44 |
| % germination | 100% | 0 | 0 | — |
| Amylase | 111 | 111 | 32 | <5 |
| β-Glucosidase | 103 | 46 | 0 | 0 |
| Catalase | 160 | 20 | 0 | — |
| Peroxidase | 75 | 9 | <5 | <5 |
| Polyphenol oxidase | 73 | 17 | 15 | 11 |

all chemical or histochemical methods devised to test viability are only partially satisfactory. Such tests can only check for one definite reaction which may to some extent be correlated with the eventual ability of the seed to germinate. Most of these tests are based on the activity of certain oxidizing enzymes. The best correlation has been found to the activity of enzymes reacting with redox dyes, such as tetrazolium, but even here positive results do not always indicate 100 per cent germination of the seed population. An X-ray contrast technique has also been used quite successfully to predict seed viability (Kamra, 1964). General chemical changes in seed composition, as viability is lost, are discussed by Owen (1956) and Abdul Baki and Andersen (1972).

Viability is retained for very long periods of time, especially in seeds having a hard seed coat, as in the Leguminosae where viability is often retained for several decades. The most extreme case of retention of viability is the case of *Nelumbo nucifera,* the Indian lotus. Seeds of this variety have been shown by radio carbon dating to be around 1,000 years old. However, carbon dating of seeds is open to many serious errors, and must be treated with great reserve (Godwin, 1968). The *Nelumbo* seeds were found in the mud of a lake bed in Manchuria and germinated after their seed coats were broken. Conservative estimates date the seeds as being about 250–400 years old. In contrast all the claims of viability of grains found in the Egyptian pyramids have been shown to be spurious.

The very long life span of many species of wild plants must be contrasted with the relatively short one of many cultivated species. In these, from a practical point of view, a high germination percentage is important and this is usually retained for relatively short periods. Some data are given in Table 3.4.

Table 3.4—Changes in Percentage Germination of Seeds During Storage
(Compiled from various sources)

| | Years of storage in which seeds germinate | | |
|---|---|---|---|
| | 70–100% | 30–60% | less than 30% |
| Wheat | 9 | — | 13 |
| Rye | 7 | — | 12 |
| Barley | 8 | — | 12 |
| Oats | 11 | 12 | -- |
| Melon | 11 | | |
| Cucumber | | 9 | |
| Spinach | | | 5 |
| Tobacco | | | 11 |
| Sunflower | | | 9 |
| Buckwheat | | | 8 |
| Alfalfa | | 11 | |
| Clover (red) | | 4 | |
| Clover (white) | | 2 | |
| Peas | 3 | | |
| Timothy grass | | 5 | |
| Carrots | 1 | 7 | 15 |
| Eggplant | 5 | 7 | 10 |
| Lettuce | 3 | — | 5 |
| Onion | — | 1 | 3 |
| Pepper | | 1 | 5 |
| Tomato | 7 | — | 10 |
| Flax | | 18 | |
| Radish | | 10 | |

It must be stressed that seed viability is not only a function of seed storage. A variety of factors to which the parent plant is exposed during seed formation and ripening can also profoundly affect subsequent viability of the seeds, after dispersion or harvest. Such factors include water supply, temperature, mineral nutrition, and light. However, these environmental factors are secondary in importance, compared to the genetic control of seed viability.

## II. External Factors Affecting Germination

In order that a seed can germinate, it must be placed in environmental conditions favourable to this process. Among the conditions required are an adequate supply of water, a suitable temperature and composition of the gases in the atmosphere, as well as light for certain seeds. The requirement for these conditions varies according to the species and variety and is determined both by the conditions which prevailed during seed formation and even more by hereditary factors. Frequently it appears that there is some correlation between the environmental requirement for germination and the ecological conditions occurring in the habitat of the plant and the seeds. In the following these various factors will be considered in detail.

### 1. *Water*

The first process which occurs during germination is the uptake of water by the seed. This uptake is due to the process of imbibition. The extent to which imbibition occurs is determined by three factors, the composition of the seed, the permeability of the seed coat or fruit to water, and the availability of water in liquid or gaseous form in the environment. Imbibition is a physical process which is related to the properties of colloids. It is in no way related to the viability of the seeds and occurs equally in live seeds and in seeds which have been killed by heat or by some other means. During imbibition molecules of solvent enter the substance which is swelling, causing solvation of the colloid particles and, in addition, occupying the free capillary spaces and the intermicellar spaces of the colloid. The swelling of the colloid results in the production of considerable pressures, called *imbibition pressure*. This is usually measured and also defined as that pressure which must be applied to the system to prevent the colloid from swelling. The imbibition pressure developed by seeds may reach hundreds of atmospheres and in colloids such as agar or gelatine pressures of many hundreds of atmospheres have been measured. The imbibition pressure is of great importance in the process of germination as it may lead to the breaking of the seed coat and also to some extent makes room in the soil for the developing seedling. The magnitude of the imbibition pressure is also an indication of the water retaining power of the seed and therefore determines the amount of water available for rehydrating the seed tissues during germination. In seeds we are dealing with the imbibition of water by hydrophilic colloids. Colloids are characterized by the size of the particles in the dispersion phase, and imbibition is a property of colloids which are in the form of a gel, i.e. where the colloidal particles constitute a more or less continuous micellar network, showing a certain amount of rigidity.

In hydrophilic gels the positive and negative charges are organized in a definite fashion, due to the presence of an electric double layer. This is made up of a very narrow fixed layer of solvent held to the surface of the solid and a second diffuse layer which extends into the bulk of the solvent. The difference of potential between the fixed layer and the freely mobile layer, is termed the *zeta potential*. Due to the zeta potential on the one hand and the dipole nature of the water molecules on the other hand electrostatic forces play an important part during swelling of hydrophilic gels. Imbibition is also accompanied by the liberation of heat, particularly during the initial absorption of water. This is an indication that true compound formation occurs during the early stages of imbibition. The volume of the imbibing substance increases during imbibition, but the volume of the hydrated colloid is smaller than that of the sum of the solvent imbibed and the colloid, before imbibition occurred. This volume change is explained by the loss of a component of translational mobility of the system due to the sorption of water during the early stages of imbibition (Glasstone, 1946, Jirgensons, 1958).

The gels occurring in nature, and in seeds in particular, are usually polyelectrolytes and contain a large number of ionic groups. The molecules themselves are of considerable molecular weight and are therefore not freely mobile. It is therefore possible to treat the imbibition of water into such polyelectrolytes according to the theories of a Donnan equilibrium. This treatment has led many authors to the view that imbibition as a whole should be treated as a special case of osmosis and that the driving forces involved are in fact the same as those concerned in osmosis. These studies consider that part of the swelling colloid, due to its immobility, acts as a semi-permeable membrane and the bulk of it as the osmotic system (Haurowitz, 1950; Katchalski, 1954).

In seeds the chief component which imbibes water is the protein. However, other components also swell. The mucilages of various kinds will contribute to swelling, as will part of the cellulose and the pectic substances. Starch on the other hand does not add to the total swelling of the seeds, even when large amounts of starch are present. Starch only swells at very acid pH or after treatment with high temperatures, conditions which do not occur in nature. The swelling of seeds (Table 3.5) therefore to some extent reflects the storage materials present in the seeds. Seed swelling depends on the pH of the solution but does not strictly follow the behaviour

Table 3.5—Imbibition by Various Seeds
Imbibition at 28°C expressed as percentage of the original weight of the seeds
(After Levari, 1960)

| Time in hours | Lettuce | Wheat | Sunflower | Vicia sativa (vetch) | Zea mays |
|---|---|---|---|---|---|
| 1 | 170 | 114 | 124 | 112 | 111 |
| 2 | 185 | 120 | 137 | 143 | 116 |
| 4 | 197 | 127 | 147 | 168 | — |
| 6 | 201 | 133 | 153 | 181 | 124 |
| 10 | 213 | 140 | 154 | 182 | — |
| 16 | 225 | — | — | — | 136 |
| 24 | 237 | 151 | — | — | 137 |
| 32 | 252 | 155 | — | — | — |
| 40 | 254 | — | — | — | — |
| 48 | 270 | 161 | — | — | — |

expected if only ampholytes were swelling. Proteins being "Zwitter-ions" show a minimum of imbibition at their isoelectric point, the imbibition rising with pH on either side of this point. Other colloids show a dependence of imbibition on pH which can be related to their dissociation constant. For agar, maximal dissociation is near neutrality and maximal swelling also occurs around pH 7·0. Imbibition is dependent on temperature and proceeds more rapidly at higher temperatures.

Shull (1920) compared the imbibition of *Xanthium* seeds having a semi-permeable membrane and split peas without such a membrane. He noted that swelling of the seeds was essentially similar to that of colloids. The $Q_{10}$ values of imbibition, in both cases, was between 1·5 and 1·8. Shull concluded that no chemical change was involved in the effect of temperature on imbibition and that it was not markedly affected by the presence of a semi-permeable membrane.

The effect of temperature on imbibition is probably complex. The viscosity of water decreases with increased temperature and its kinetic energy increases. The kinetic energy is directly proportional to the absolute temperature, while the molecular velocity varies as the square root of the absolute temperature.

Modern views on osmotically-induced flow of water assume that the bulk of the flow is hydrodynamic mass flow (Pappenheimer, 1953) through the pores of the membrane and not diffusion. As mentioned above, imbibition may be considered as a special case of osmosis. Therefore any effect of temperature on the structure of the colloid and the dimensions of its intermicellar spaces might affect the rate of imbibition. The final volume obtained by seeds imbibing at low temperature is greater than that resulting from the rapid imbibition at higher temperature (Fig. 3.1). However, experimental evidence shows that for seeds these differences are very slight.

The composition of the germination medium also determines the imbibition of seeds, as it determines the availability of water. This is of significance under natural conditions where the solution in which the seeds are found is not usually pure water. As the concentration of solutes in the solution increases, imbibition decreases. This is largely due to osmotic effects. In addition, however, a direct effect of the ions on the seed is also frequently observed. Toxic effects may be present, for example under very saline conditions.

The preceding discussion considered the problem of imbibition by the colloid materials in the seed. The entry of water into seeds is, however, determined in the first instance by the permeability of the seed coat or the fruit coat. Seeds which are surrounded by an impermeable seed coat will not swell even under otherwise favourable conditions. Impermeable seed coats are frequently found in Leguminosae as well as in other groups. The seed coat is usually a multi-layered membrane containing a number of layers or cells. Frequently it shows selective permeability toward certain substances. The impermeability of the seed coat, or its selective permeability, is frequently the cause of dormancy (see Chapter 4). Various external factors can cause changes in the permeability of the seed coat. For example, heat-killed seeds often imbibe water more rapidly than the corresponding viable seeds, probably because the permeability of the seed coat is increased by the heat treatment. Denny (1917) showed that the seed coats of various species showed

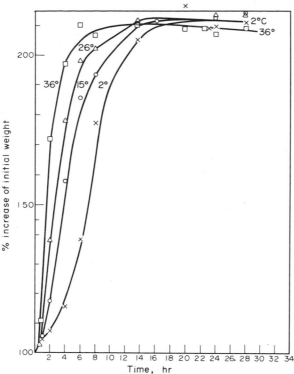

Fig. 3.1. Imbibition of heat-killed peas at different temperatures.

×——× 2°    △——△ 26°
□——□ 36°    ○——○ 15°

different permeability to water. He was able to relate these differences to the composition of the seed coat and especially to lipoid components (Table 3.6). Permeability of the seed coat is generally greatest near the micropylar end of the seed, where it is almost invariably thinner than the rest of the seed coat. However, in some cases the micropyle does not contribute to permeability. In *Quercus* species the pericarp is normally impermeable and unless it is damaged water penetrates through the capscar (Bonner, 1968). The presence of mucilages in the seed coat improves the ability to imbibe water and the artificial addition of mucilages has essentially the same effect. Mucilage reduces the sensitivity of the seed to soil water tension (Harper and Benton, 1966). The ability of seeds to absorb water from the soil as compared to water uptake from solution is determined not only by the osmotic potential of the soil solution, but also by the matric potential of the soil. Contact of the seed surface with soil particles is very important in this respect. Although the water potential of seeds is usually very low, compared to that of the soil, seeds can also lose water to the air, which also has a low water potential. The final uptake of water will then be determined by the relative values of these water potentials. Normally only the water potential in the immediate vicinity of the seeds will determine their imbibition. Various attempts have been made to relate the imbibition

Table 3.6—Effect of Extracting Seed Coats with Hot Water or Hot Alcohol on
the Water Permeability of the Isolated Seed Coat
(Compiled from data of Denny, 1917)

| Seed | Solvent | %<br>Increase in<br>permeability | Probable seed coat<br>constituent restrict-<br>ing water permeability |
|------|---------|-----------|------------------|
| *Arachis hypogea* | Hot water | 170 | ⎫ Tannins |
| *Arachis hypogea* | Hot alcohol | 80 | ⎬ Lipids (2% of seed<br>⎭ coat) |
| *Prunus amygdalus* | Hot water | 500 | ⎱ Pectic substances |
| *Prunus amygdalus* | Hot alcohol | 350 | ⎰ Lipids |
| *Citrus grandis* | Hot water | 0 | ⎫ Fatty substances |
| *Citrus grandis* | Hot alcohol | 0 | ⎬ surrounded by thick<br>⎭ pectinized walls |
| *Cucurbita maxima* | Hot water | 0 | ⎱ |
| *Cucurbita maxima* | Hot alcohol | 700 | ⎰ Lipids |

of seeds to temperature and seed quality. Blacklow (1972) developed an equation for
the imbibition of *Zea*:

$$\frac{\mathrm{d}W}{\mathrm{d}t} = K\left(f(t) - W\right) + b$$

where $K$ is a measure of the permeability of the seeds to water during the
experimental phase of water uptake, $W$ the water content of the seeds, $f(t)$ the water
capacity of the seeds and $b$ a measure of the linear phase of water uptake. With the
aid of this mathematical model Blacklow was able to predict the imbibition of seeds
under various conditions.

## 2. Gases

Germination is a process related to living cells and requires an expenditure of
energy by these cells. The energy-requirement of living cells is usually sustained by
processes of oxidation, in the presence or absence of oxygen, i.e. respiration or
fermentation. These involve an exchange of gases, an output of carbon dioxide in
both cases and also the uptake of oxygen in the case of respiration. Consequently
seed germination is markedly affected by the composition of the ambient atmosphere. Most seeds germinate in air, i.e. in an atmosphere containing 20 per cent oxygen
and a low percentage, 0·03 per cent, of carbon dioxide. However, many authors have
shown that certain seeds respond to an increase in the oxygen tension above 20 per
cent by increased germination, e.g. *Xanthium* and certain cereals.

Thornton (1934) showed that the germination of many cultivated seeds, if
maintained at 20 per cent oxygen, was unaffected by increased concentration of
carbon dioxide. Seeds of *Daucus carota* and *Rumex crispus* respond to increased
oxygen concentrations in the dark by increased germination; no germination at 20 per
cent $O_2$, 3 per cent at 40 per cent $O_2$, and 24 per cent at 80 per cent $O_2$ (Gardner, 1921).

Most seeds will show lower germination if the oxygen tension is decreased appreciably below that normally present in the atmosphere. Although it is usually assumed that rice germinates well under anaerobic conditions, caused by flooding of the seeds, work in China has thrown doubt on this. Tang, Wang and Chih (1959), and Chu and Tang (1959), showed that anaerobic conditions lead to the formation of abnormal seedlings and that these abnormalities are prevented by the presence of oxygen. In contrast, a number of seeds show increased germination as the oxygen content of the air is decreased below 20 per cent. The best established cases seem to be those of *Typha latifolia* and *Cynodon dactylon*, which germinate better in the presence of about 8 per cent oxygen than in air (Morinaga, 1926). It is important to note that quite different results are obtained if the dilution of the air is done with nitrogen or with hydrogen. For example, Morinaga found that while white clover, *Trifolium repens*, seeds give 52 per cent germination in air, they give only 47 per cent in air diluted with 60 per cent nitrogen, but give 70 per cent germination in the presence of air equally diluted with hydrogen. Clearly hydrogen has an effect on germination. In none of these cases is anything known about the oxygen tension within the seeds. Some species germinate as well in 2 per cent $O_2$ as in air, e.g. *Celosia, Portulaca* and cucumber (Siegel and Rosen, 1962). Lettuce seeds germinated 58 per cent in an atmosphere of $N_2$—95 per cent and $O_2$ 5 per cent; 79 per cent when the $O_2$ was raised to 15 per cent and 96 per cent in 80 per cent $N_2$ and 20 per cent $O_2$. Seedling growth was, however, depressed even at 15 per cent $O_2$ (Harel and Mayer, 1963). At least in one case—*Trifolium subterraneum*—pre-incubation of seeds in the absence of oxygen led to higher subsequent germination in air (Ballard and Grant Lipp, 1969). This case shows not a reduced requirement for $O_2$ but rather a case of dormany breaking. A similar requirement for anaerobiosis in the dark part of the inductive cycle has been demonstrated for *Eragrostis ferruginea* (Fujii, 1963). Dormancy is often ascribed to the impermeability of the seed coat to certain gases. Such differential permeability of seed coats was demonstrated by Brown (1940) (see Chapter 4). The determining factor which may be involved in the effect of oxygen on germination will be considered in Chapter 4 on dormancy, as well as in Chapter 7.

The effect of carbon dioxide is usually the reverse of that of oxygen. Most seeds fail to germinate if the carbon dioxide tension is greatly increased, as shown, for example, by Kidd (1914), for *Hordeum vulgare* and *Brassica alba*. However, in some cases at least, there appears to be a minimal requirement for carbon dioxide in order that germination can occur. This seems to be so for *Atriplex halimus* and *Salsola*, as well as for lettuce. Other *Atriplex* species are resistant to high carbon dioxide concentrations, as shown by Beadle (1952) and as found for cultivated seeds, provided that $O_2$ concentration is kept constant. Very high carbon dioxide concentrations, which prevent germination, seem to have a favourable effect on the keeping of seeds. Kidd (1914) showed that the life-span of *Hevea brasiliensis* seeds was prolonged by sealing the seeds in an atmosphere containing 40–45 per cent $CO_2$. However, an atmosphere of nitrogen was even more effective. For lettuce and onion seeds, storage under carbon dioxide decreased the number of chromosomal aberrations which occur in the mitoses during the germination, subsequent to storage (Harrison and McLeish, 1954). Cases are known where increases in $CO_2$ concentration increase germination, e.g. *Phleum pratense* (Maier, 1933). A dormancy breaking by $CO_2$ has been demonstrated for *Trifolium subterraneum* seeds. Generally 2·5 per cent was optimal and imbibition occurred only when the $CO_2$ concentration reached 10–15 per cent. The $CO_2$ effect was, as might be expected, temperature dependent (Ballard, 1967). For carbon dioxide also, the determining factor probably is the internal concentration. This is determined by the permeability of the seed coat to carbon dioxide accumulating within the seed (see also dormancy).

Small amounts of ethylene have long been known to promote the germination of some seeds. This effect has been ascribed to dormancy breaking or to effects on the rate of growth immediately after the onset of germination (Esashi and Leopold, 1968). Others claim its effect is directly on germination. Certainly ethylene production begins very soon after the onset of imbibition (Abeles and Lonski, 1969). Ethylene interacts with growth hormones and its effect must be regarded as that of a growth regulator. In lettuce it promotes germination in seeds treated with either red or far-red light but not in the dark.

It might be supposed that because of the sensitivity of seeds to gases, pressure would have a marked effect on germination. However, pressures up to 200 atmospheres seem to have little effect on germination of seeds (Vidaver, 1972).


## 3. *Temperature*

Different seeds have different temperature ranges within which they germinate. At very low temperatures and very high temperatures the germination of all seeds is prevented. The precise sensitivity is very different according to the species. A rise in temperature does not necessarily cause an increase in either the rate of germination or in its percentage. Germination as a whole is therefore not characterized by a simple temperature coefficient. This can be understood if it is appreciated that germination is a complex process and a change in temperature will affect each constituent step individually, so that the effect of temperature which is observed will merely reflect the overall resultant effect.

In studying the effect of temperature on germination one must distinguish the resistance of dry seeds to various temperatures and the effect of temperature on the actual germination. Many dry seeds are fairly resistant to extreme temperatures. Thus it has been shown that the viability of seeds is not affected by placing them at the temperature of liquid air. High temperatures up to about 90°C for prolonged periods of time are tolerated by many seeds such as radish, turnips and poppy, as shown by David (1936) who studied oil-containing seeds. In other cases the temperature tolerance is lower. Above 90°C the resistance of the seeds is greatly lowered. There probably is some correlation between the storage materials in the seed and their heat resistance, although seeds may still be viable after treatment at high temperature; the subsequent development of the seedling is often adversely affected (Levitt, 1956).

In the range of temperatures within which a certain seed germinates there is usually an optimal temperature, below and above which germination is delayed but not prevented. The optimal temperature may be taken to be that at which the highest percentage of germination is attained in the shortest time. The minimal and maximal temperatures for germination are the highest and lowest temperatures at which germination will occur. The maximal temperature at which germination still occurs may be as high as 48°, for example, for *Cucumis sativa* (Knapp, 1967).

The minimal temperature is frequently ill defined because germination is so slow that the experiments are often terminated before germination could in fact have occurred. In many cases it is stated that the optimal temperature for germination shifts with length of the germination period. In these cases the term "optimal" is used ambiguously and in fact refers to the optimal temperature when germination is observed after some definite, arbitrary, time interval. If a different time interval is chosen, then a different temperature may be optimal according to this usage of the term (Fig. 3.2).

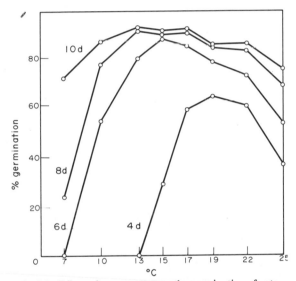

Fig. 3.2. Effect of temperature on the germination of oats.
(After Edwards, 1932 and Attenberg, 1928)
The percentage germination as determined at the different temperatures after various periods
of time, viz. 4, 6, 8 and 10 days.

The temperatures at which germination can still occur can be modified. Thus in lettuce, abscisic acid, although slowing down the rate of germination, reduced the temperature at which the seeds failed to germinate by some 3–4° while kinetin raised the temperatures by as much as 10° (Reynolds and Thompson, 1971). Thus even the range of temperatures at which germination can occur is not absolute and can be modified by exogenous compounds. It is reasonable to assume that the endogenous growth substance levels also play a role in determining the temperature range of germination.

The temperature at which different seeds germinate and the range within which they germinate, is determined also by the source of the seeds, genetic differences within a given species, e.g. varietal differences, as well as age of the seeds. A few examples of

Table 3.7—Temperature Ranges in which Germination Occurs for
Different Seeds
(Compiled from various sources)

| Seeds | Temperature | | |
|---|---|---|---|
| | Minimum | Optimum | Maximum |
| *Zea mays* | 8–10 | 32–35 | 40–44 |
| *Oryza sativa* | 10–12 | 30–37 | 40–42 |
| *Triticum sativum* | 3–5 | 15–31 | 30–43 |
| *Hordeum sativum* | 3–5 | 19–27 | 30–40 |
| *Secale cereale* | 3–5 | 25–31 | 30–40 |
| *Avena sativa* | 3–5 | 25–31 | 30–40 |
| *Fagopyrum esculentum* | 3–5 | 25–31 | 35–45 |
| *Cucumis melo* | 16–19 | 30–40 | 45–50 |
| *Convolvulus arvensis* | 0·5–3 | 20–35 | 35–40 |
| *Lepidium draba* | 0·5–3 | 20–35 | 35–40 |
| *Solanum carolinense* | 20 | 20–35 | 35–40 |
| *Nicotiana tabacum* (Florida cigar wrapper) | 10 | 24 | 30 |
| *Delphinium* (annual) | — | 15 | 20–25 |

ranges of temperature in which germination occurs are given in Table 3.7, but in view of the preceding discussion such data must be treated with great reserve.

In contrast to those seeds which germinate readily if held at one specific temperature, instances are known where a periodic alternation of temperature is required for germination to occur, as in *Oenothera biennis, Rumex crispus, Cynodon dactylon, Nicotiana tabacum, Holcus lanatus, Agrostis alba, Poa trivialis* and many others. The most usual cases of such alternations are diurnal ones, between a low and a high temperature. An examination of some of the results obtained with alternating temperatures suggests that what is determining germination is the actual alternation of temperature. The temperatures chosen, between which the seeds were alternated, appear to be of secondary importance, provided they are in a range, within which the seeds can germinate and their viability is not affected. For example *Agrostis alba* seeds gave 69 per cent germination if alternated between 12° and 21°C, and 95 per cent for an alternation between 21° and 28°C or 21° and 35°C. At constant temperature the germination was 49 per cent at 12°C, 53 per cent at 21°C, 72 per cent at 28°C and 79 per cent at 35°C (Lehman and Aichele, 1931).

This effect of alternating temperatures on germination was analysed by Cohen (1958) for lettuce seeds, by measuring the actual temperature reached by the seeds. He concluded that neither the rate nor the duration of the temperature change is determining germination. He suggests that the actual change of temperature of the seeds themselves is the determining factor.

Stimulation by alternating temperatures has been variously ascribed to an effect of temperature on sequential reactions during germination or to mechanical changes occurring in the seed.

The results of Cohen indicate that neither of these interpretations is acceptable and that the change takes place in a macro-molecular structure in the seed, which in its original form prevents germination in some way.

*Lycopus europaeus* seeds have an absolute requirement for fluctuating temperature, but there was no evidence for a critical temperature requirement. Successive temperature cycles were cumulative in their effect on germination (Thompson, 1969). This would be consistent with the view that structural changes are involved.

The effect of temperature on germination is not independent of other factors. Thus instances of interdependence between temperature and light are known for celery, *Amaranthus* and other seeds, where light promotes germination at unfavourable high temperatures, but not at low ones.

*Physalis franchetti* seeds germinate well in the dark between 5 and 15°C. Between 15°C and 35°C they require light for their germination and the light requirement increases with the temperature (Baar, 1912). A very similar situation exists for lettuce seeds.

The interaction of temperature and light is also demonstrated by work on *Rumex crispus* (Taylorsen and Hendricks, 1972). These seeds will germinate when exposed to a temperature shift from 20° to 35° for short periods. High temperature, 30°C, reduces germination. Exposure to 30° during imbibition reduces the subsequent response to temperature shifts. This effect of high temperature is reversed by exposure to low energies of red light (R). It appears that germination in the dark, caused by the temperature shift, is due to an effect of the temperature shift on pre-existing phytochrome in the seeds in the $P_{FR}$ form.

Temperature effects are also known in relation to after-ripening of seeds, as well as to secondary dormancy. This point will be considered later.

## 4. *Light*

Among cultivated plants there is very little evidence for light as a factor influencing germination. The seeds of most cultivated plants usually germinate equally well in the dark and in the light. In contrast, among other plants much variability in the behaviour toward light is observed. Seeds may be divided into those which germinate only in the dark, those which germinate only in continuous light, those which germinate after being given a brief illumination and those which are indifferent to the presence or absence of light during germination. Daily illuminations have also been shown to affect germination, the effects being similar to those of photoperiodism in flowering.

The importance of light as a factor in the germination of seeds has long been recognized. Already at the end of the nineteenth and the beginning of the twentieth century, a number of papers by Cieslar (1883), Gassner (1915) and by Lehman (1913) analysed some of these phenomena. It is possible that the light sensitivity of seeds has some relation to their germination in their natural habitat although such a view is contested by others (Niethammer, 1922). Under natural conditions seeds may be shed so as to fall on the soil or enter the soil or be covered by leaf litter, thus creating different conditions of light during germination. Among the species which have been investigated for their light response during germination at least half showed a light requirement. For example, Kinzel (1926) lists hundreds of plant species which he divided into several categories. His first groups include those germinating at or above 20°C in the light (about 270 species) and in the dark (114 species). Two further categories germinate in the light (190 species) and in the dark (81 species) after severe frost, and others showed similar behaviour after mild frost (52 species in the light and 32 species in the dark). A group of seeds which are indifferent to light or dark includes 33 species. A selection from his data is given in Table 3.8.

Table 3.8—Response of Seeds of Different Species to Light
(From data of Kinzel, 1926)

A—seeds whose germination is favoured by light
B—seeds whose germination is favoured by dark
C—seeds indifferent to light or dark

| A | B | C |
|---|---|---|
| *Adonis vernalis* | *Ailanthus glandulosa* | *Anemone nemorosa* |
| *Alisma plantago* | *Aloe variegata* | |
| *Bellis perennis* | *Cistus radiatus* | *Bryonia alba* |
| *Capparis spinosa* | *Delphinium elatum* | *Cytisus nigricans* |
| *Colchicum autumnale* | *Ephedera helvetica* | |
| *Erodium cicutarium* | *Evonymus japonica* | *Datura stramonium* |
| *Fagus silvatica* | *Forsythia suspensa* | *Hyacinthus candicans* |
| *Genista tinctoria* | *Gladiolus communis* | |
| *Helianthemum cha-* | | |
| maecistus | *Hedera helix* | *Juncus tenagea* |
| *Iris pseudacorus* | *Linnaea borealis* | *Linaria cymbalaria* |
| *Juncus tenuis* | *Mirabilis jalapa* | |
| *Lactuca scariola* | *Nigella damascena* | *Origanum majorana* |
| *Magnolia grandiflora* | *Phacelia tanacetifolia* | *Pelargonium zonale* |
| *Nasturtium officinale* | *Ranunculus crenatus* | *Sorghum halepense* |
| *Oenothera biennis* | *Silene conica* | *Theobroma cacao* |
| *Panicum capillare* | *Tamus communis* | *Tragopogon pratensis* |
| *Resedea lutea* | *Tulipa gesneriana* | *Vesicaria viscosa* |
| *Salvia pratense* | *Yucca aloipholia* | |
| *Suaeda maritima* | | |
| *Tamarix germanica* | | |
| *Taraxacum officinale* | | |
| *Veronica arvensis* | | |

Such a classification of light requirement is probably an oversimplification, as light requirement varies during storage. In some species a light requirement only exists immediately after harvesting (e.g. in *Salvia pratensis, Saxifraga caespitosa* and *Epilobium angustifolia*), while in other species this effect persists at least for a year (e.g. *Epilobium parviflorum, Salvia verticillata* and *Apium graveolens*), while in yet other species it only develops during storage.

The question of light requirement has been the subject of detailed studies under laboratory conditions. These studies have shown, as is to be expected, that different spectral zones affect germination quite differently. The early work on the effect of light distinguished only between fairly wide spectral bands. These showed that light below 290 nm inhibited germination in all seeds tested. Between 290 nm and 400 nm no clear-cut effects on germination were detected. In the visible range, 400 nm–700 nm, it was shown that light in the range 560 nm–700 nm and especially red light, usually promoted germination, while blue light was said to inhibit. Flint and McAllister (1935 and 1937) determined these spectral ranges more accurately for lettuce seeds and showed that the most effective light in promoting germination is that having a wavelength of 670 nm and that a germination inhibiting zone of the spectrum has its maximal activity at 760 nm. This latter was far more effective than the inhibiting zone in the blue region of the spectrum.

Kincaid (1935) studied the effect of light on the germination of tobacco seeds. He found that as little as 0·01 seconds of sunlight was effective in stimulating germination and even moonlight could stimulate. Although the action spectrum was not studied in detail, a filter transmitting between 435–580 nm with a maximum at 544 nm was very effective. The seed coat had an absorption maximum at 510 nm. Resuehr (1939) studied the germination of *Amaranthus caudatus*; the action spectrum showed three inhibitory zones, at about 450 nm, between 475 and 490 nm and between 700 and 750 nm. Two promoting zones were observed at 640 nm, where strong stimulation was obtained and at 675–680 nm, where slight stimulation was noted. For the usually light-inhibited seeds of *Phacelia*, Resuehr observed light stimulation of germination at 640 nm, and five inhibitory zones, at 350 nm, 450 nm, 475–490 nm, 680–690 nm and above 1000 nm. Orange light has been reported to stimulate germination of *Physalis franchetti* (Baar, 1912). It was more effective than blue-violet light.

Whether or not blue light does in fact inhibit germination or stimulate it, was disputed for many years, because it was contended that the light used by Flint and McAllister contained far-red or infra-red light as well as blue light. However, later both Wareing and Black (1957 and 1958) and Evenari, Neuman and Stein (1957) showed that blue light can in fact inhibit germination. The latter showed that under certain conditions blue light may also stimulate germination. Whether germination was stimulated or inhibited depended entirely on the exact period of illumination as related to the beginning of imbibition.

The sensitivity of seeds to light increases with time of imbibition. Maximum sensitivity is reached after about 1 hour and does not exactly coincide with the completion of imbibition (cf. Table 3.5 and Fig. 3.3). Even storage of seeds at high relative humidities is sometimes sufficient to make them light-sensitive. If seeds are given a light stimulus while imbibed and then dried, the stimulatory effect is retained (Gassner, 1915; Kincaid, 1935). The precise time at which the seeds reach maximum sensitivity for any given species has been a matter of dispute. There have also been differences of opinion whether the sensitivity reaches a maximal value and then drops again or whether it remains at a high level. It appears that these differences can be related to the light intensity or amount of radiation at which sensitivity was tested on the one hand, and to the time interval which was allowed to elapse between the illumination and the determination of germination percentage on the other hand. Figure 3.3 shows some of these effects for lettuce seeds. After 10 hours or more of imbibition 30 seconds of illumination were not enough to induce maximal germination.

In tobacco seeds Kincaid (1935) found that high light intensities were effective after short periods of imbibition (several hours), while low light intensities were most effective after 4 days. After 10 days of imbibition the seeds no longer responded to illumination.

The effect of light is being studied on an increasing number of plant species. Some of the earlier detailed work was concerned with lettuce (*Lactuca sativa*), *Lepidium*

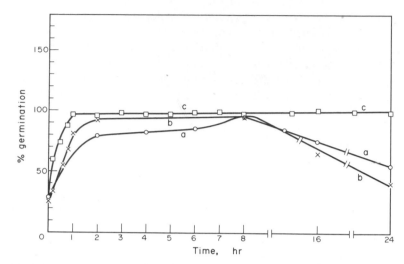

Fig. 3.3. Effect of length of imbibition period, prior to illumination, on light sensitivity of
lettuce seeds, var. Grand Rapids, to red light at 26°C.
(a) ○——○ 30 sec light, total incubation time 48 hours
(Reconstructed from Evenari and Neuman, 1953)
(b) ×——× 30 sec light, total incubation time 72 hours
(c) □——□ 2 min light, total incubation time 72 hours
(b), (c) after Poljakoff-Mayber and Lang, unpublished)

*virginicum, Nicotiana tabacum* and various *Amaranthus* species. For lettuce and
*Lepidium* the precise spectral peaks for germination, stimulation and inhibition, by
short illuminations, have been redetermined and earlier results essentially confirmed,
showing stimulation at 670 nm and inhibition at 730 nm. An important new finding was
the reversibility of both germination stimulation and germination inhibition by
alternating illuminations as shown by Borthwick *et al.* (1952). Lettuce seeds whose
germination is stimulated by red light can be inhibited if they are subsequently
illuminated with infrared (or far-red) light. Further illumination with red light will
again induce germination. These effects are in fact very similar to, if not identical with,
those also known for flowering, etiolation, unfolding of the plumular hook of bean
seedlings as well as pigment formation in certain fruit and leaves.

All these phenomena show a very similar action spectrum. The action spectrum is
that associated with the plant pigment phytochrome (P), which was deduced from the
biological response of the various tissues to illumination. The action spectra for the
germination of lettuce and *Lepidium* seeds are shown in Fig. 3.4. Butler *et al.* (1959)
were able to show that there are changes in the absorption spectrum of an intact tissue,
*Zea mays* coleoptiles, as a result of illumination with red or far-red light. This is
illustrated in Fig. 3.5.

The stimulation of germination by red light, and its inhibition by far-red light, can be
repeated many times and always the nature of the last illumination decides the
germination response. As in all cases where light-effects are observed, so here also,
the effect noted is dependent on both the intensity of the light, i.e. its energy, and the

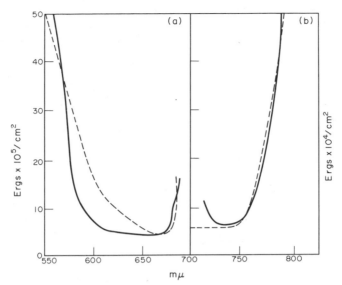

Fig. 3.4. Action spectrum for germination promotion (a) and inhibition (b) of *Lactuca sativa* and *Lepidium virginicum* seeds to 50%.
(After Toole *et al.*, 1956)
The curves show the amount of radiant energy at different wave lengths required to promote or inhibit germination to half its maximum value.
----- Lettuce, in (a) scale × 0.04, in
(b) scale × 10
——— *Lepidium*

duration of the illumination. Provided that all the light is absorbed, then, within certain ranges, the response of germination will be a function of the product of light intensity and duration of illumination, i.e. a function of the total energy of irradiation. This applies both to germination stimulation and inhibition, as well as to all the other responses to red and far-red irradiation. In order to show the reversibility of red and far-red light effects, suitable intensities and durations must be selected. The range in which these observations have been made is for total irradiations of $2 \times 10^4$ ergs/cm$^2$ for promotion in lettuce and $1 \cdot 4 \times 10^6$ ergs/cm$^2$ for promotion in *Lepidium* and $6 \times 10^5$ and $3 \times 10^4$ ergs/cm$^2$ for inhibition respectively in these species (see also Fig. 3.4). The period of irradiation is between a few seconds and half an hour or more.

The presence of phytochrome in seeds is now adequately documented. Not only has the presence of phytochrome been demonstrated, its transition from one form to another has been proved. An example for this is shown in Fig. 3.6. From this it can be seen that the $\Delta(\Delta\text{O.D.})$ at 670 nm of the entire seeds increases with length of imbibition, and a smaller but similar effect is observed at 730 nm. The detailed interpretation of the effect of light on germination is immensely complicated. In order to explain the effect of light of a given intensity and duration we must take into account the absorption spectrum of phytochrome (Fig. 3.7), which shows clearly the overlap in absorption between the R and FR forms of phytochrome as well as its absorbance in the blue. In addition we must consider the transformation which phytochrome can

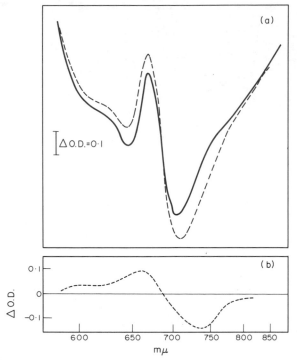

Fig. 3.5. Absorption spectra of maize coleoptiles following red or far-red irradiation (a) as well as the difference spectrum (b) (far-red irradiated spectrum minus red irradiated spectrum).
(Butler *et al.*, 1959)
—— Red irradiated
----- Far-red irradiated

undergo in seeds (Fig. 3.8). These transformations are in essence the following:

(1) Activation of phytochrome from an inactive preexisting form.
(2) Synthesis of phytochrome from its precursors.
(3) Transformation of $P_R$ to $P_{FR}$ by red light, or less effectively by blue light.
(4) Transformation of $P_{FR}$ to $P_R$ by far-red light, which requires higher energies than the reverse conversion.
(5) The reversion in the dark of $P_{FR}$ to $P_R$.
(6) The inverse dark reversion, detected in dry seeds, causing a partial conversion of $P_R$ to $P_{FR}$ even in the dark.
(7) The destruction, enzymatic or otherwise of $P_{FR}$.
(8) The reaction of $P_{FR}$ with some unknown substance in the seed such as a chemical compound or a membrane which results in the actual effect on germination.

Transformations 5 and 6 take place via a number of intermediates, some of which are quite stable, (Kendrick and Spruit, 1974).

From this it is clear that phytochrome is normally in a state of equilibrium between its two forms $P_R$ and $P_{FR}$ and that many factors will determine the actual ratio of the

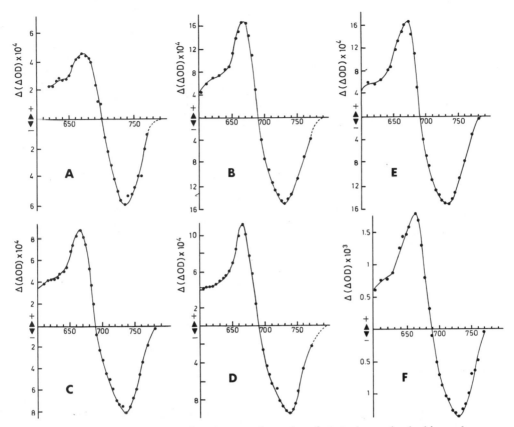

Fig. 3.6. Difference spectra for phototransformation of phytochrome in gherkin seeds.
    A: dry seeds
    B: after 24$^h$ imbibition
    C: after 3$^h$ imbibition
    D: after 8$^h$ imbibition
    E: after 16$^h$ imbibition
    Sample thickness: 4 mm
    Temperature of measurement: 0°C
    Temperature of imbibition: 22°C
    F: Difference spectrum for phototransformation of phytochrome on a single
    pumpkin embryo after 24 h imbibition at 22°C
    Temperature of measurement: 22°C
                   (After Malcoste *et al.*, 1970)

two forms at a given time. Furthermore there is at least some evidence to indicate that there is more than one pool of phytochrome in plant tissues. It is not yet clear what are the relative sizes of such pools and how and to what extent they are involved in the light responses in germination. In general it can be stated that germination is determined by the amount of $P_{FR}$ as per cent of the total phytochrome in seed. However, the percentage required in order to induce germination seems to be quite variable, depending on the seed involved. In lettuce the amount required for germination may be as high as 30–40 per cent, while in *Eragrostis* the amount required

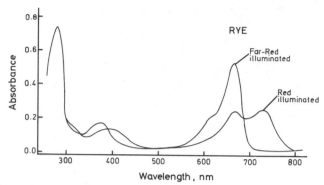

Fig. 3.7. Absorption spectra of purified rye phytochrome, specific activity 0·720, $A_{280}/A_{665}$ 1·27, in 0·1M sodium phosphate buffer, pH 7·8, 5% (v/v) glycerol, 4°C, after 4-min R and FR irradiation.
(After Rice *et al.*, 1973)

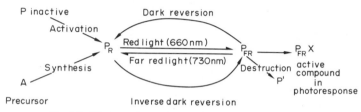

Fig. 3.8. The transformations undergone by phytochrome.

for germination seems to be much smaller. In *Nigella* the $P_{FR}$ level required is as low as 1–3 per cent and in tomato seeds 22 per cent. While the situation is thus reasonably well understood for those seeds whose germination is induced by red light and in which the red light effect is reversed by far-red light, other cases of light effects are less well understood. As already mentioned promotion of germination by blue light is known. This can best be regarded as a typical phytochrome effect, since as already mentioned phytochrome absorbs light in the blue part of the spectrum.

The interpretation of inhibition of germination by light is more complex. It has been shown that both in the case of *Phacelia tanacetifolia* and *Nemophila insignis* germination is inhibited by white light. The effect of white light is due to its components of blue and far-red light. The far-red component has a typical phytochrome action spectrum. The effect of far-red light in these cases is confined to a certain period after seed imbibition (Table 3.9). According to Rollin (1968) the seeds of *Nemophila* germinate when the $P_{FR}/P_{total}$ ratio lies above 40 per cent or below about 20 per cent. Minimum germination occurs when $P_{FR}/P_{total}$ is 28 per cent while in the case of *Phacelia* minimal germination occurs at a value of 10 per cent. These effects are still readily explained on the basis of the effect of FR light converting the phytochrome from the $P_{FR}$ to $P_R$, causing inhibition, unless all the P is in the $P_R$ form. The effect of blue light in inhibiting germination is even more complex. Generally it is still supposed that here too phytochrome is involved. However, there have been suggestions that an

Table 3.9—Germination of *Nemophila insignis* in
Response to Various Light Regimes
(After Rollin & Maignan, 1967)

|                                      | % Germination |
| ------------------------------------ | ------------- |
| Dark                                 | 90            |
| 20 hr R + dark                       | 82            |
| 20 hr dark + 16 hr FR + dark         | 15            |
| 20 hr R + 16 hr FR + dark            | 68            |
| 20 hr dark + 16 hr R + dark          | 79            |

Red light 670 nm, 110 $\mu$w/cm$^2$
FR light 713 nm, 120 $\mu$w/cm$^2$

additional pigment might be involved. This will have to be established much more firmly before it can be accepted.

The last major effect of light which must be considered is the so-called High Energy Reaction or HER. There are seeds in which germination is inhibited by prolonged FR. Although much doubt existed about the nature of this HER it is now believed to be mediated via phytochrome. The effect of light on germination, in detail, is exceedingly complicated and the literature is vast, but some recent reviews on the subject are by Black, 1969, Briggs, 1972 and Rollin, 1973.

It must be mentioned that the response of seeds to light is modified by various internal and external factors. Osmotic stress, the presence of growth promoters or growth inhibitors, the $O_2$ tension and so on, all can change the duration and intensities of light required to evoke a certain response. Even the spectrum of radiation to which the parent plant has been exposed during seed formation, as demonstrated, for example, for seeds of *Arabidopsis thaliana*, affects the response of the seeds to light. In general it may be said that all internal and external factors affecting germination, interact in some way.

The way in which the active form of phytochrome brings about its effect on germination is still quite unclear. The identity of the hypothetical compound X, with which the active form combines, is also unknown. Numerous suggestions have been made. It has been proposed that the phytochrome acts in some way on cell membranes or that it is involved in gene activation. Almost all forms of metabolism have been shown to be altered due to phytochrome transformation. However, the present state of knowledge does not give a definite answer to the question: which of these effects are the cause of stimulated (or inhibited) germination and which are the results. Until this question is resolved it will also be impossible to satisfactorily interpret the interaction of the effect of light with numerous other factors controlling germination.

Another type of radiation which affects living organisms is shortwave irradiation such as $\gamma$-rays and X-rays. This type of radiation can also affect germination and development of seedlings. Such radiation affects not only germination but also chromosomal structure and integrity and therefore also affects seedling development and even subsequent seed formation. However, a consideration of the effects of this type of radiation as well as that of $\alpha$ and $\beta$-irradiation and the irradiation with other particles, is outside the scope of this book.

Strictly speaking, the effects of light discussed here are effects on the breaking of dormancy or its induction. Other light effects, perhaps more directly related to the phenomenon of dormancy, are the photoperiodic response of the germination of certain seeds as well as the complex interactions between light and other factors affecting germination. These will be dealt with in the following chapter.

## Bibliography

Abdul-Baki, A. A. and Anderson, J. D. (1972) *Seed Biology*, vol. II, p. 283 (ed. Kozlowski T. T.) Acad. Press, N.Y.

Abeles, F. B. and Lonski, J. (1969) *Plant Physiol.* **44**, 277.

Attenberg, A. (1928) *Landw. Versuchsstation* **67**, 129.

Baar, H. (1912) *S. Acad. Wiss. Math.-Naturn. K.L.* **121**, 667.

Ballard, L. A. T. (1967) In *Physiologie Ökologie und Biochemie der Keimung*, p. 209 (ed. H. Boriss), Ernst-Moritz-Arndt-Universität: Greifswald.

Ballard, L. A. T. and Grant-Lipp, A. E. (1969) *Austr. J. Biol. Sci.* **22**, 279.

Barton, L. V. (1961) *Seed Preservation and Longevity*, Leonard Hill, London.

Beadle, N. C. E. (1952) *Ecology*, **33**, 49.

Beal, W. J., as cited by Shull, G. M. (1914) *The Plant World* **17**, 329.

Becquerel, P. (1932) *C.R. Acad. Sci. Paris* **194**, 2158.

Becquerel, P. (1934) *C.R. Acad. Sci. Paris* **199**, 1662.

Black, M. (1969) *SEB Symposium*, **23**, 193.

Blacklow, W. M. (1972) *Crop Science* **12**, 643.

Bonner, F. T. (1968) *Bot. Gaz.* **129**, 83.

Borthwick, H. A., Hendricks, S. B., Parker, M. W., Toole, E. H. and Toole, V. K. (1952) *Proc. Nat. Acad. Sci., Wash.* **38**, 662.

Briggs, W. R. and Rice, H. V. (1972) *Ann. Rev. Plant Physiol.* **23**, 293.

Brown, R. (1940) *Ann. Bot. N.S.* **4**, 379.

Butler, W. L., Norris, K. H., Siegelman, H. W. and Hendricks, S. B. (1959) *Proc. Nat. Acad. Sci. U.S.* **45**, 1703.

Ching, T. M., Parker, M. C. and Hill, D. D. (1959) *Agron. J.* **51**, 680.

Chu, G. and Tang, P. S. (1959) *Acta Bot. Sinica* **8**, 212.

Cieslar, A. (1883) *Forsch. Gebiete Agrikultur Physik* **6**, 270.

Cohen, D. (1958) *Bull. Res. Council, Israel* **6D**, 111.

Crocker, W. (1938) *Bot. Rev.* **4**, 235.

David, R. L. (1936) *Influence des Températures Elevées sur la Vitalité des Graines Oléagineuses*, Imprimerie Universitaire, Aix en Provence.

Denny, F. E. (1917) *Bot. Gaz.* **63**, 373.

Denny, F. E. (1917) *Bot. Gaz.* **63**, 468.

Discussions of the Faraday Society No. 27 (1959) *Energy Transfer with Special Reference to Biological Systems*. Aberdeen University Press, Scotland.

Duvel, J. W. T. (1905) *U.S.D.A. Bureau of Plant Industry Bull.* 83.

Edwards, T. I. (1932) *Quart. Rev. Biol.* **7**, 428.

Esashi, Y. and Leopold, A. G. (1968) *Plant Physiol.* **43**, 871.

Evenari, M. and Neuman, G. (1953) *Bull. Res. Council, Israel* **3**, 136.

Evenari, M. and Stein, G. (1953) *Experientia* **9**, 94.

Evenari, M., Neuman, G. and Stein, G. (1957) *Nature Lond.* **100**, 609.

Flint, H. L. and McAllister, E. D. (1937) *Smithsonian Miscellaneous Collection* **96**, No. 2.

Flint, H. L. and McAllister, E. D. (1935) *Smithsonian Miscellaneous Collection* **94**, No. 5.

Fujii, T. (1963) *Plant & Cell Physiol.* **4**, 357.

Gardner, W. A. (1921) *Bot. Gaz.* **71**, 249.

Gassner, G. (1915) *Z. Bot.* **7**, 609.

Glasstone, S. (1946) *Textbook of Physical Chemistry*, Macmillan.

Godwin, H. (1968) *Nature* **220**, 708.

Goss, W. L. (1924) *J. Agr. Res.* **30**, 349.

Griffiths, A. E. (1942) *Cornell Univ. Agr. Exp. Stat. Memoir*, 245.

Harel, E. and Mayer, A. M. (1963) *Physiol. Plant.* **16**, 804.

Harper, J. L. and Benton, R. A. (1966) *J. Ecol.* **54**, 151.
Harrison, B. J. and McLeish, J. (1954) *Nature*, Lond. **173**, 593.
Haurovitz, F. (1950) *Chemistry and Biology of Proteins*, Academic Press, N.Y.
Hendricks, S. B., Toole, E. H., Toole, V. K. and Borthwick, H. A. (1959) *Bot. Gaz.* **121**, 1.
Holden, M. (1959) *J. Sci. Food. and Agr.* **12**, 691.
Jirgensons, B. (1958) *Organic Colloids*, Elsevier, Amsterdam.
Kamra, S. K. (1964) *Proc. Intern. Seed Test Assoc.* **29**, 519.
Katchalski, A. (1954) *Progr. in Biophysics*, **14**, 1.
Kendrick, R. E. and Spruit, C. J. B. (1974) *Planta* **120**, 265.
Kidd, F. (1914) *Proc. Roy. Soc. B.* **87**, 410.
Kincaid, R. R. (1935) *Tech. Bull.* **277**, Univ. Florida, Agr. Exp. St.
Kinzel, W. (1926) *Frost und Licht, Neue Tabellen*, Eugen Ulmer, Stuttgart.
Kivilaan, A. and Bandurski, R. S. (1973) *Am. J. Bot.* **60**, 140.
Knapp, R. (1967) *Flora* **157b**, 3.
Lehmann, E. (1913) *Biochem. Z.* **50**, 388.
Lehmann, E. and Aichele, F. (1931) *Keimungsphysiologie der Gräser*, Verlag Ferdinand Enke, Stuttgart.
Levari, R. (1960) Ph.D. thesis, Jerusalem.
Levitt, J. (1956) *The Hardiness of Plants*, Acad. Press, New York.
Maier, W. (1933) *Jahrb. Wiss. Bot.* **78**, 1.
Malcoste, R., Boisard, J., Spruit, C. J. P. and Rollin, P. (1970) *Ned. Landbouwhogeschool, Wageningen* **70-16**, 1.
Morinaga, T. (1926) *Amer. J. Bot.* **13**, 126.
Morinaga, T. (1926) *Amer. J. Bot.* **13**, 159.
Niethammer, A. (1957) *Biochem. Zschr.* **185**, 205.
Owen, E. Biasutti (1956) The Storage of Seeds for Maintenance of Viability, *C.A.B. Bulletin* **43**, Commonwealth Bureau of Pastures and Field Crops.
Pappenheimer, J. R. (1953) *Physiol. Rev.* **33**, 387.
Resuehr, B. (1939) *Planta* **30**, 471.
Reynolds, T. and Thompson, P. A. (1971) *Physiol. Plant.* **24**, 544.
Rice, H. V., Briggs, W. R. and Jackson-White, C. J. (1973) *Plant Physiol.* **51**, 917.
Roberts, E. H. (ed.) (1972) *Viability of seeds*, Chapman & Hall, London.
Rollin, P. (1968) *Bull. Soc. Franc. Physiol. Veget.* **14**, 47.
Rollin, P. (1972) In *Phytochrome*, p. 230. (ed. W. Shropshire Jr. and K. Mitrakos), Acad. Press, New York.
Rollin, P. and Maignan, G. (1967) *Nature* **214**, 741.
Shull, G. H. (1914) *The Plant World* **17**, 329.
Shull, C. A. (1920) *Bot. Gaz.* **69**, 361.
Siegel, S. M. and Rosen, L. A. (1962) *Physiol. Plant.* **15**, 437.
Tang, P. S., Wang, F. C. and Chih, F. C. (1959) *Acta Bot. Sinica* **8**, 199.
Taylorson, R. B. and Hendricks, S. B. (1972) *Plant Physiol.* **50**, 645.
Thompson, P. A. (1969) *J. Exp. Bot.* **20**, 1.
Thornton, N. C. (1943–45) *Contr. Boyce Thompson Inst.* **13**, 355.
Toole, E. H. and Brown, E. (1946) *J. Agr. Res.* **72**, 201.
Toole, E. H., Toole, V. K., Borthwick, H. A. and Hendricks, S. B. (1955) *Plant Phys.* **30**, 15.
Toole, E. H., Hendricks, S. B., Borthwick, H. A. and Toole, V. K. (1956) *Ann. Rev. Plant. Physiol.* **7**, 299.
Turner, J. H. (1933) Bull. Misc. Information **6**, p. 257. Royal Botanic Gdns. Kew.
Vidaver, W. E. (1972) *SEB Symposium* **26**, 159.
Villiers, T. A. (1974) *Plant Physiol.* **53**, 875.
Wareing, P. F. and Black, M. (1957) *Nature, Lond.* **180**, 385.
Wareing, P. F. and Black, M. (1958) Nature, Lond. **181**, 1420.
Went, F. W. and Munz, P. A. (1949) *El. Aliso* **2**, 63.

Chapter 4

# DORMANCY, GERMINATION INHIBITION AND STIMULATION

Many seeds do not germinate when placed under conditions which are normally regarded as favourable to germination, namely an adequate water supply, a suitable temperature and the normal composition of the atmosphere. Such seeds can be shown to be viable, as they can be induced to germinate by various special treatments. Such seeds are said to be dormant or to be in a state of dormancy. Dormancy can be due to various causes. It may be due to the immaturity of the embryo, impermeability of the seed coat to water or to gases, prevention of embryo development due to mechanical causes, special requirements for temperature or light, or the presence of substances inhibiting germination.

Attempts have been made to classify states of dormancy (Nikolaeva, 1969): In these attempts of classification exogenous and endogenous causes of dormancy are distinguished. These are further divided according to the supposed type of dormancy, e.g. physical, chemical, etc. However, it is fairly obvious today that more than one cause may be responsible for the dormancy of a given seed. We will therefore not follow a systematic classification of dormancy but try to understand some of the underlying mechanisms.

In very many species of plants the seeds, when shed from the parent plant, will not germinate. Such seeds will germinate under natural conditions, if they are kept for a certain period of time. These seeds are said to require a period of *after-ripening*.

After-ripening may be defined as any changes which occur in seeds during storage as a result of which germination is improved. This is the most generally used definition of this term. An alternative definition is also possible. This would define after-ripening as those processes which must occur in the embryo and which can occur only with time and which cannot be caused by any known means other than suitable storage conditions of the seeds.

After-ripening often occurs during dry storage. In other cases storage of the seeds in the dry state does not cause after-ripening. The seeds must be stored in the imbibed state, usually at low temperatures, in order that they after-ripen. This is termed *stratification* and will be dealt with later.

There is much diversity in the conditions under which after-ripening in "dry storage" occurs. The length of the storage period required is also variable. Thus some barley varieties after-ripen after a fortnight, while *Cyperus* after-ripens only over a period of 7 years. After-ripening during dry storage is difficult to classify. In cereals germination is low at harvest and increases during storage. In lettuce, *Amaranthus* and *Rumex* the fresh seeds can germinate but the requirements for their germination are very specific. These special requirements tend to disappear during storage. For example, fresh lettuce seeds only germinate below 20°C but after storage germination occurs even at 30°C. The reverse case obtains in *Amaranthus retroflexus*. Fresh seeds only germinate above 30°C, while stored seeds germinate over a wider range down to 20°C. The light requirement in lettuce also disappears during prolonged dry storage. This type of response to storage should not, strictly speaking, be termed "after-ripening", although this term is often applied to these phenomena.

The necessity for a period of after-ripening may be due to a number of factors. Various kinds of change may consequently occur during this process. In the case of an immature embryo further anatomical and morphological changes may occur. In other seeds chemical changes must occur in the seed before it can germinate. Frequently germination of such seeds can be forced by suitable treatment although the resulting seedlings may be abnormal.

Immature embryos are known among many families of plants, although they are most commonly associated with plants which are saprophytic, parasitic or symbiotic. Such embryos may attain full maturity either during the actual process of germination or they may mature as a preliminary to germination. In either case the changes only occur if the seeds are kept under conditions favourable to germination, and differentiation between the two types is extremely difficult. Among the plants in which immature embryos occur are the Orchidaceae and Orobanchaceae, as well as some *Ranunculus* species. The period required for such embryos to reach maturity varies from a few days to several months. In all these, processes of differentiation at the anatomical and morphological level occur during the period of after-ripening. In other seeds, such as those of *Fraxinus*, the embryo, although morphologically complete, must still increase its size before germination can take place.

In contrast, in many seeds no visible anatomical or morphological changes occur in the embryo during after-ripening. In these it must be assumed that the process of after-ripening is the result of chemical or physical changes which occur within the seed or seed coat. The composition of the storage materials present in the seed may alter, the permeability of the seed coat may change, substances promoting germination may appear or inhibitory ones may disappear. In no case has it been possible to ascribe after-ripening to any one definite event.

## I. Secondary Dormancy

In contrast to those seeds which fail to germinate when shed, but germinate after a period of after-ripening, other seeds will germinate readily under favourable conditions. However, these seeds may lose their readiness to germinate. This phenomenon is called *secondary dormancy*. Secondary dormancy may develop spontaneously in seeds due to changes occurring in them, as in some species of *Taxus* and *Fraxinus*. These changes may be the reverse of those described as occurring during after-ripening. Sometimes secondary dormancy is induced if the seeds are given all the conditions required for germination except one. For instance, the failure to give light to light-requiring seeds, or illuminating light-inhibited seeds such as

*Nigella* and *Phacelia*, induces dormancy. *Phacelia* seeds will not germinate if kept continuously in the light. When they are returned to, and kept in, the dark for long periods they will eventually germinate provided the period in the light has not been too prolonged. If the seeds are kept for very long in the light they are then unable to germinate in the dark under conditions under which they previously germinated (Mato, 1924).

Too high or too low temperatures for germination may also induce secondary dormancy, for example in *Ambrosia trifida* and *Lactuca sativa*. Lettuce seeds if kept imbibed in the dark at high temperatures will not germinate even if returned to lower temperatures, and no longer respond to light. Only much more drastic treatment such as a chilling or chemical treatment with gibberellin will induce germination.

In *Eragrostis* dormancy is induced by cycling the seeds between a low and a higher temperature during imbibition, the low temperature being the determining factor. A light requirement was noted after the induction of secondary dormancy (Isikawa *et al.*, 1961).

Low oxygen tension will cause dormancy in *Xanthium* (Davis, 1930). High carbon dioxide tensions may cause secondary dormancy, e.g. in *Brassica alba* (Kidd, 1914). It may also be induced by chemical treatment of the seeds, e.g. the induction of light sensitivity in light-indifferent varieties of lettuce seeds by treatment with coumarin. This latter phenomenon is closely related to the problem of germination inhibitors and will be dealt with later.

It may be assumed that the mechanisms underlying secondary dormancy are the same as those of dormancy in general. The fact that new requirements for germination arise indicates that, while the seed is exposed to conditions which impose secondary dormancy, metabolic changes occur. These may be changes in permeability, shifts in the balance of the forms of phytochrome, changes in inhibitor and stimulator relationships or other metabolic events.

## II. Possible Causes of Dormancy

The possible causes of dormancy have already been listed. In the following these will be considered in greater detail. At the same time, the means by which seed dormancy may be broken will be discussed. Dormancy breaking is frequently of economic importance. The mechanism of artificial dormancy breaking and the natural process leading to the same effect are frequently similar.

### 1. *Permeability of Seed Coats*

A very widespread cause of seed dormancy is the presence of a hard seed coat. Such hard seed coats are met with in many plant families and usually can cause dormancy in one of three ways. A hard seed coat may be impermeable to water, impermeable to gases or it may mechanically constrain the embryo.

The impermeability of seed coats to water is most wide-spread in the Leguminosae. The seed coats of many members of this family are very hard, resistant to abrasion and covered with a wax-like layer. Such seed coats appear to be entirely impermeable to water. In some cases the entry of water into the seed is controlled by a small opening in the seed coat, which is closed with a cork-like filling consisting of suberin—the strophiolar cleft with the strophiolar plug. Only if this plug is removed or loosened in some way, will water enter the seeds. This mechanism was first described by Hamly (1932) who also investigated various artificial ways of dormancy-breaking in such seeds. He was able to show that vigorous shaking of the seeds had the effect of loosening or removing the strophiolar plug, thus rendering the seeds permeable to water. This treatment is frequently called impaction and has been applied to seeds of *Melilotus alba*, *Trigonella arabica* and *Crotalaria aegyptiaca*. The opening of the strophiolar gap may be reversible. The majority of seeds showing a strophiolar cleft belong to the Papilionaceae.

Other hard-coated seeds do not possess a strophiolar cleft. Such seeds become permeable to water only if the seed coat is abraded in some way. In nature, the seed coat may be broken down or punctured by mechanical abrasion, microbial attack, passage through the digestive tract of animals or exposure to alternating high and low temperatures which, by expanding and contracting the seed coat cause it to crack. Under laboratory conditions and in agriculture other means of rendering the seed coat permeable have been adopted. These are either shaking with some abrasive, to cause mechanical breakage, or chemical treatment. The chemical treatment is chiefly of two kinds: removal of the waxy layer of the seed coat by some suitable solvent such as alcohol, or treatment with acids (Table 4.1). The mechanism by which the latter act is

Table 4.1—Effect of Acid Treatment on Germination of Various
Seeds
(From data of Wycherley, 1960; Khudairi, 1956)

|  | % Germination | |
| --- | --- | --- |
|  | Untreated | Treated with sulphuric acid |
| *Calopogonium muconoides* | 30 | 65 |
| *Centrosema pubescens* | 30 | 70 |
| *Flemingia congesta* | 45 | 60 |
| *Pueraria phaseoloides* | 20 | 70 |
| *Purshia tridentata* | 1 | 84 |
| *Cercocarpus montanus* | 1 | 52 |
| *Atriplex canescens* | 11 | 16 |

far from clear. Possibly chemical decomposition of seed coat components is involved, which may be analogous to the breakdown processes occurring during microbial attack or passage through the digestive tract. The treatment with alcohol is especially effective for members of the family Caesalpiniaceae. In all the cases studied, the various treatments could be shown directly to increase the water uptake of the seeds and the number of seeds which swelled (Table 4.2). The dormancy-breaking action or abrasion, alcohol or sulphuric acid could therefore be directly related to an increase in the permeability of the seeds to water. Nevertheless such vigorous treatment of seeds

Table 4.2—Changes in Permeability of Seeds to Water Caused by Various Treatments
The change in permeability is expressed as the percentage of seeds which swell
(From data of Barton and Crocker, 1948, and Koller, 1954)

| Species | No treatment | Alcohol treatment | Impaction | Sulphuric acid | Mechanical abrasion |
|---------|-----------|-----------|-----------|-----------|-----------|
| *Trigonella arabica* | 5 | 4 | 100 | — | 96 |
| *Crotalaria aegyptiaca* | 0 | 20 | 80 | 95 | 95 |
| *Melilotus alba* | 1 | 0 | 86 | — | 100 |
| *Cassia artemisioides* | 2 | 57 | 3 | — | 100 |
| *Parkinsonia aculeata* | 2 | 100 | 8 | — | 100 |

is likely to induce other changes also, such as permeability to gases, changes in sensitivity to light or temperature and possibly even destruction or removal of inhibitory substances. Thus, although there is no doubt that increase in water permeability is an essential part of the dormancy-breaking action of such treatments, it is by no means certain that it is the only result of these treatments.

The germination of seeds is determined not only by their ability to imbibe water, but also by the conditions during imbibition. Thus in barley water uptake at low temperatures from the liquid phase leads to subsequent damage to the seedling, but this is not the case if the seeds are exposed to water vapour. Excess water often leads to dormancy or poor germination. This need not necessarily be due to seed coat effects on water permeability. In the case of *Blepharis*, a vital role is played by the mucilages in the seed coat. In the presence of excess water the mucilage becomes a diffusion barrier to oxygen (Gutterman *et al.*, 1967). Excess water may also encourage the development of large mixed populations of micro-organisms in and around the seed envelope, which compete with the embryo for available oxygen.

Frequently, seed coats are impermeable to gases despite the fact that the seeds are permeable to water. This impermeability may be either toward carbon dioxide or oxygen or both. That such differential permeability exists seems fairly well established, despite the apparently small differences in molecular diameter of the substances involved. Only very few artificial membranes are known which show such differential permeability.

The most frequently cited case of impermeability to oxygen is provided by *Xanthium*. In *Xanthium* Crocker (1906) showed that the fruit contains two seeds, an upper and a lower, which differ in their ability to germinate. Later both Shull (1911 and 1914) and Thornton (1935) showed that they differ in their requirement for external oxygen pressure for germination. The upper seeds seem to require a much higher oxygen concentration than the lower ones, to give 100 per cent germination. The upper seed gives 100 per cent germination at 21°C only in pure oxygen, while the lower one only requires 6 per cent oxygen for full germination. The oxygen requirement of both seeds is reduced by higher temperatures (Thornton, 1935). Intact seeds needed higher oxygen concentrations for germination than excised embryos from both upper and lower seeds, which give 100 per cent germination at 1·5 and 0·6 per cent oxygen respectively. This indicates that the seed envelope is impermeable to oxygen. It is not clear whether the impermeability of the seed envelope reduces the internal oxygen concentration to such an extent that oxidative respiratory mechanisms are reduced or whether some other mechanism is operative. The latter has been suggested by Wareing and Foda (1957) who indicated that the high oxygen requirement of the upper seeds is due to the presence of an inhibitor which has to be destroyed by oxidation before germination can occur. Recently, Porter and Wareing (1974) showed that there is no difference in oxygen permeability of the seed coat of imbibed upper and lower seed of *Xanthium*, although there is a three fold difference in the permeability of the dry seed. Even under these conditions diffusion of oxygen is adequate for respiration. They suggest that the high oxygen requirement is caused by the oxidation of an inhibitor.

Improved germination by increased oxygen tension is also shown by wild oats, indicating restricted permeability to oxygen. Spaeth (1932) reported that the nucellar membrane restricted the oxygen supply of seeds of *Tilia americana* and also advanced evidence that the testa is impermeable to moisture. In neither case is the mechanism clear. A striking instance of the complexity of the function of a permeability barrier to oxygen is that of *Sinapis arvense* (Edwards, 1969, 1972). Seed coats of *S. arvense* are permeable to water but less so to oxygen. Removal of the seed coat allows an adequate supply of oxygen to reach the embryo. However, it was found that the requirement for oxygen by the embryo did not exceed the supply. In the presence of low oxygen tensions there was production of a germination inhibitor which tended to accumulate within the seed and which might prevent germination. Raised oxygen tensions slowed down inhibitor formation and permitted normal germination (Table 4.3).

Table 4.3—Effect of $O_2$ Concentration on $O_2$
Uptake and Growth Inhibitor Production by
*Sinapis* Embryos
(After Edwards, 1969)

| $O_2$ concn (atmosp.) | $O_2$ uptake ml/g/4 hr | Inhibitor content (units) |
|---|---|---|
| 0 | 0 | 11·8 |
| 0·05 | 1·0 | 10·5 |
| 0·1 | 2·0 | 9·2 |
| 0·2 | 2·5 | 8·8 |
| 1·0 | 2·5 | 8·5 |

Seed coat permeability to $O_2$ has also been directly measured in the case of apple seed. Imbibition of the seeds at 20° resulted in a rapid decrease in their permeability to $O_2$ while at 4°, $O_2$ permeated the seed coat quite readily. It seems possible that the seed integuments actually used up oxygen and thereby limited its supply to the embryo (Come, 1968). The permeability of seed coats can be changed by the presence of external factors. For example, saponins which arise in the soil from decomposition of plant material affect the permeability of the seed coats of lucerne to $O_2$ (Marchaim *et al.*, 1972). It seems quite likely that this is not an isolated instance. Changes in seed coat permeability to water were induced in *Pisum elatius* by changing the conditions of drying the seeds after harvest. In the absence of $O_2$ during drying of the seeds they were fully permeable to water, while in the presence of $O_2$ impermeable seeds were obtained (Marbach and Mayer, 1974). These results indicate that the permeability of seed coats can and does change in response to environmental factors.

Instances are known where seed coats are differentially permeable to oxygen and to carbon dioxide. Brown (1940) has shown that the nucellar membrane of *Cucurbita pepo* shows a differential permeability to oxygen and to carbon dioxide. The isolated inner membrane is more permeable to carbon dioxide, $15 \cdot 5 \text{ ml/cm}^2$ hr, than to oxygen, $4 \cdot 3 \text{ ml/cm}^2$ hr. Although the inner membrane is more permeable to gases than the outer one, it nevertheless is the inner membrane which in fact controls permeability. The permeability of the membrane to carbon dioxide is increased by treatment with chloroform or heat. This treatment is supposed to kill the living cells and thus to change the structure and permeability of the inner membrane. It did not affect the permeability to oxygen, indicating that the two gases followed different paths of diffusion through the membrane. The significance of this for the germination of the Cucurbitaceae is not clear.

In *Xanthium*, carbon dioxide at high concentrations cannot induce dormancy of the intact seeds in the presence of oxygen. As little as 1 per cent oxygen was sufficient to reduce the effectiveness of carbon dioxide in inducing dormancy. The induction process was temperature-dependent (Thornton, 1935).

Carbon dioxide does not invariably induce dormancy. In Chapter 3, a number of instances of improved germination have already been mentioned which are due to the presence of low concentrations of carbon dioxide. A more extreme case is that of the dormancy-breaking action of carbon dioxide on *Trifolium subterraneum* (Ballard, 1958). Carbon dioxide at concentrations between $0 \cdot 3$ and $4 \cdot 5$ per cent was effective. Carbon dioxide above 5 per cent was found to have an inhibitory effect. Treatment of the seed with activated carbon also stimulated germination and the results were consistent with the supposition that this treatment, too, resulted in a raising of the carbon dioxide concentration. These findings have been extended to other species of *Trifolium*, as well as to *Medicago* and *Trigonella* species. Carbon dioxide treatment caused breaking of dormancy in those cases where cold treatment was effective and even in some cases where cold treatment was ineffective, sometimes only after prolonged storage. Neither carbon dioxide nor cold broke the dormancy of the freshly-harvested seeds (Grant-Lipp and Ballard, 1959). The germination of *Phleum pratense* is stimulated by raised concentrations of $CO_2$, both in the light and in the dark, even after removal of the seed coat (Maier, 1933).

Axentjev (1930) studied the effect of seed coats on the germination of seeds of many species. In many seeds dormancy was partly caused by seed coat effects, probably due to the impermeability of the seed coat, and partly by a light requirement, and these factors are to some extent additive. Thus, *Cucumis melo* is light-inhibited, but its germination in the dark is further improved by removal of the seed coat (see Table 4.4).

Seed coats may also exert a physical restraint on the developing embryo. If the thrust developed during imbibition and growth is inadequate, the seed will not rupture the seed coat and will fail to germinate. Thus certain kinds of dormancy are caused by the inability of the embryo to develop the necessary thrust. This appears to be the case

Table 4.4—Effect of Light and Seed Coats on Germination of Various Seeds
Results given as % Germination
(Compiled from Axentjev, 1930)

| Seed | Treatment | Temp. | 5 | | 7 | | 8 | | 10 | | 12 | | 14 | |
|---|---|---|---|---|---|---|---|---|---|---|---|---|---|---|
| | | | Light | Dark | Light | Dark | Light | Dark | Light | Dark | Light | Dark | Light | Dark |
| *Cucumis melo* | Whole seed | 17·5 to 20°C | 0 | 41 | 4 | 65 | 5 | 72 | 20 | 78 | 30 | 87 | 68 | 100 |
| | Without seed coat | | 0 | 89 | 15 | 96 | 25 | 98 | 48 | 100 | 60 | 100 | 80 | 100 |
| | Cotyledons pierced, seed coat removed | | 19 | 84 | 32 | 98 | 36 | 100 | | | 42 | 100 | 48 | 100 |
| | Root pierced with seed coat removed | | 6 | 72 | 18 | 93 | 18 | 97 | | | 23 | 100 | 27 | 100 |
| *Nigella arvensis* | Whole seed | 17·5 to 19°C | 5 | 38 | 7 | 47 | — | 48 | | | | | | |
| | Seed coat pierced | | 2 | 18 | 6 | 48 | 8 | 51 | | | | | | |
| *Oenothera* | Whole seed | 14 to 20°C | 0 | 0 | 0 | — | 0 | — | 0 | 0 | | | 3 | — |
| | Seed coat pierced | | 1 | 1 | 13 | 5 | 16 | 7 | 24 | 9 | | | 34 | 12 |
| *Phacelia tanaceti-folia* | Whole seed | 7·5 to 10°C | | | 22 | 92 | 22 | — | 53 | — | | | | |
| | Seed pierced | | | | 49 | 100 | 51 | — | | | | | | |
| | Whole seed | 11·5 to 13°C | | | 14 | 78 | | | | | | | | |
| | Seed pierced | | | | 69 | 100 | | | | | | | | |

for the dormant seeds of *Xanthium* (Esashi and Leopold, 1968). In such cases the mere mechanical weakening of the seed coat will relieve dormancy. In other cases it is possible that there is a requirement for chemical dissolution of the seed coat, by enzymes produced by the radicle. However, evidence for this latter mechanism is still not completely convincing (Ikuma and Thiman, 1963).

## 2. Temperature Requirements

The effect of temperature on germination as such has already been discussed. An additional effect of temperature must be considered. Many seeds require an exposure to some definite temperature before they are placed at the temperature favourable to germination. Seeds are treated at either high or low temperatures which do not permit germination. They will germinate only on transfer to some other temperature. To respond to this treatment the seeds must be fully imbibed. Seeds requiring such temperature treatment must be considered as being dormant and events occurring in them are akin to after-ripening in wet storage. The most commonly known and used procedure of exposure of the seeds to low temperature under moist conditions is termed *stratification*. Evidently, during stratification changes take place within the seeds. Embryo growth has been noted during stratification. Such growth has been described in cherry seeds by Pollock and Olney (1959). These authors were able to show that during after-ripening in moist storage at 5°C the embryonic axis increased in cell number, dry weight and total length. An increased oxygen uptake was also noted in the embryonic axis and the leaf primordium, but not in the whole seed. On a cellular basis, an increase of oxygen uptake was especially marked in the embryonic axis. It appears that during after-ripening the energy supply to the embryo is increased. Investigations of the changes in nitrogen and phosphate content of cherry seeds during stratification were also carried out (Olney and Pollock, 1960). Nitrogen content rose in the embryonic axis and in the leaf primordium but the nitrogen content per cell remained constant. Phosphorus content, however, increased both on the basis of the total per organ and per cell. The phosphorus and nitrogen which appear in the embryo presumably originate in the storage organs. Some of these results are shown in Table 4.5. All the changes were either absent or much less marked if the seeds were held at 25°C instead of at 5°C. Some of the changes observed in the phosphate content of the embryo are presumably related to the greater rate of respiration in this organ.

Table 4.5—Changes Occurring in Cherry Seeds During Stratification
(Compiled from graphs of Pollock and Olney, 1959; Olney and Pollock, 1960)

| | Axis length, mm | | No. of cells/axis | | Dry Wt/axis ($\mu$g) | | $O_2$ uptake of axis as % of non-after-ripened | | Total N/axis ($\mu$g) | | Total P/axis ($\mu$g) | |
|---|---|---|---|---|---|---|---|---|---|---|---|---|
| Temp (°C) | 5° | 25° | 5° | 25° | 5° | 25° | 5° | 25° | 5° | 25° | 5° | 25° |
| Initial | 1·6 | 1·6 | 180 | 180 | 400 | 400 | 100 | 100 | 23 | 23 | 2·0 | 2·0 |
| After 8 weeks | 1·75 | 1·6 | 220 | 180 | 400 | 320 | 300 | 100 | 21 | 19 | 2·9 | 2·0 |
| After 16 weeks | 2·3 | 1·6 | 265 | 210 | 600 | 320 | 550 | 100 | 29 | 20 | 3·6 | 1·5 |

The metabolism of peach seeds during stratification has also been investigated (Flemion and de Silva, 1960). Marked changes in amino acid, organic acid, and phosphate composition were noted during after-ripening. However, it was impossible to establish a definite relationship between these changes and the actual ending of dormancy. A similar situation was found also with regard to growth substances. A later paper showed that seedlings of peach could be obtained without chilling (Flemion and Prober, 1960). Seedlings were formed if the embryos were excised and the cotyledons removed. Thus, chilling is not absolutely essential to seedling establishment from these embryos. It is possible that the seedlings were formed from a lateral bud and not from the terminal bud. This seems more likely to Flemion and de Silva than that the cotyledons are actually inhibiting embryo growth, although this possibility is not excluded by them.

A number of enzymes have also been shown to change during stratification. Thus catalase and peroxidase were shown to increase in *Sorbus aucuparia, Rhodotypos kerrioides* and *Crategus* (Flemion, 1933; Eckerson, 1913). These changes in the enzymes may be the direct cause of emergence from dormancy but it seems much more likely that they are the secondary result of other changes in the seeds. For example, Barton (1934) was able to show a complete absence of correlation between increase in catalase activity during after-ripening and the completion of the after-ripening of *Tilia* seeds.

Many Rosaceous seeds contain cyanogenic glycosides such as amygdalin. These are often broken down during stratification, liberating hydrocyanic acid. There is some coincidence in time between the cessation of the liberation of HCN and the completion of after-ripening during stratification. However, it is not known whether there is any causal relationship between the two processes.

In many cases no changes have as yet been observed during chilling treatment, but the evidence that before treatment the seeds failed to germinate and after treatment they did, indicates that internal changes nevertheless took place. In general it appears that most phases of metabolism can be affected during stratification. From the numerous investigations which have been carried out it is not possible to point to any one metabolic event which is involved in dormancy breaking by stratification.

It is sometimes possible to force seeds requiring stratification to germinate by other means, e.g. by complete removal of the seed coats or by removal of the cotyledons (Flemion and Prober, 1960). Such experiments were carried out by workers at the Boyce Thompson Research Institute, particularly on Rosaceous seed. Seeds forced to germinate in this way do not form normal seedlings, the seedlings being either dwarfed or otherwise deformed (Barton and Crocker, 1948). However, cold treatment of the seedlings will again cause normal growth. Cold treatment apparently increases the growth capacity of the embryo and seedling.

During stratification the growth substance balance in seeds is changed considerably. The presence of growth inhibitors has been reported for a number of seeds. The most widespread of these appears to be abscisic acid (Fig. 4.1) although other compounds are probably also involved. During stratification the level of abscisic acid (ABA) drops in the seed for example in *Fraxinus americana, Juglans regia* and *Corylus avellana*. In some cases the chilling requirement can be replaced by treatment of the seeds by gibberellic acid, e.g. in *Betula pubescens*. It is therefore probable that in some cases

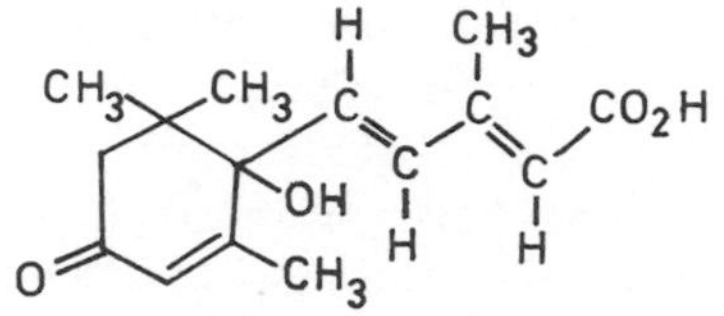

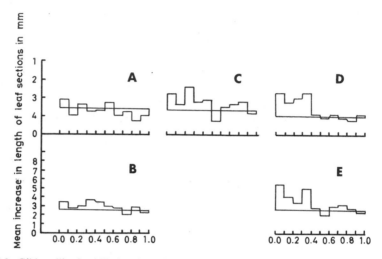

Fig. 4.1. Structure of 2-trans-abscisic acid (ABA).

Fig. 4.2. Gibberellins in chilled and unchilled hazel seeds. Dwarf maize leaf-section assays of purified extracts of 100 g of seed material following chromatography with butanol/ammonia. A and B, unchilled, C, chilled 6 weeks. D and E, chilled 12 weeks. B and E were assayed simultaneously. Least significant differences (P = 0·05) in mm: A 1·70, B 2·16, C 1·29, D 1·18, E 2·07. Increase in length of leaf sections in 0·01 mgm/l gibberellic acid: A 3·10, C 4·75, D 3·05, B and E 7·85.
(Frankland and Wareing, 1966)

the result of stratification is the formation of gibberellic acid (Fig. 4.2), possibly by initiation of its biosynthesis. In yet other instances chilling induces the formation of cytokinins. It is therefore evident that stratification acts, among other things, by shifting the balance between growth inhibitors to growth promoters, in favour of the latter. As a result germination and growth can take place. The mode of action of the growth promoters and inhibitors will be discussed in Chapter 6.

The conditions required for effective stratification are often similar to the natural conditions to which seeds are exposed. Many seeds are shed in autumn and then subjected to low temperatures during the winter months while they are exposed to moist conditions under leaf-litter or in the soil. Although this would lead one to expect that stratification requirements are similar in many plants, they in fact differ greatly. This difference exists both with regard to the length of the cold treatment and also with regard to the actual temperature during the treatment. Even in one family, e.g. the Rosaceae, and within one genus, marked differences exist. For example, Barton and Crocker (1948) cite experiments to show that while *Rosa multiflora* has a requirement of 2 months at 5–10°C, *Rosa rubiginosa* requires 6 months' stratification at 5°C, and will not respond to treatment at either 10°C or 0·5°C. However, such a precise temperature requirement is rare and usually stratification can be carried out within a fairly wide temperature range. The best known cases of seeds requiring low temperature treatment for germination are found among the Rosaceae and among various conifers. However, in other families, too, such requirements are known, as for instance in *Juglans nigra* (Juglandaceae), *Aster* (Compositae), *Adonis* and *Anemone* (Ranunculaceae), *Iris* (Iridaceae) as well as many other, including aquatic plants. In the latter the temperature treatment must be given while the seeds are actually immersed in water. Fuller lists of plants requiring stratification, giving details of time and temperatures used, are given by Crocker and Barton (1953), and by Nikolaeva (1969).

There is very little evidence in the literature to indicate that high temperatures in themselves break dormancy. Cases are known, as already mentioned, where high temperatures cause a change in the structure of the seed coats, thus causing a change in permeability. In contrast, many instances are known where there is interaction between the effects of low temperatures and somewhat raised ones. Probably a differential temperature requirement for growth of different parts of the embryo is involved here.

Crocker and Barton (1953) state that the seeds of *Paeonia suffruticosa* germinate by root growth, but the shoot fails to develop. If the germinated seeds are kept at normal temperatures, no further development occurs. However, if they are placed at a low temperature (1–10°C) for 2 or 3 months and kept moist, the bud of the epicotyl shoot develops normally when the seedlings are again transferred to normal temperatures. This phenomenon is called epicotyl dormancy by these authors. Similar phenomena occur in some species of *Lilium*.

Other cases are known where brief exposure to slightly elevated temperatures promotes subsequent germination at lower temperatures after exposure to light, for example in *Poa pratensis* and *Lepidium virginicum*. The mechanism of the effect of the raised temperatures is in no way clear. In many desert seeds, storage at 50°C promotes germination, while after storage at low temperatures germination and seedling growth is poor. The same seeds if stored at 20°C required prolonged storage and maturation, of up to 5 months, for normal germination. In this case high temperatures may be regarded as breaking dormancy. However, if the temperatures were raised to 75°C the seeds failed to germinate (Capon and Van Asdall, 1967).

In many seeds germination is promoted by alternating temperature changes, which may be either diurnal or seasonal. Diurnal temperature changes have already been discussed in Chapter 3 (See also Table 4.6). A clear-cut effect is observable for *Nicotiana* in the dark when alternated between 20 and 30°C diurnally. Smaller effects are observable for some of the other seeds when the germination at 20 or 30°C is compared with that when they are alternated between 20 and 30°C. Seasonal changes of temperatures may affect germination by affecting the actual development of the embryo or in other ways. The practice of stratification already mentioned is, in fact, analogous to one kind of seasonal change of temperature. Other combinations of temperature alternations are met with, depending on the type of seed. Stratification is

Table 4.6—Effect of Light and Temperature on the Germination of Various Seeds
(Compiled from data of Toole *et al.*, 1955, and Koller, 1954)

| Seed | Light condition | % Germination at temperature stated | | | | | |
|------|-----------------|------|------|------|------|------|---------|
|      |                 | 15° | 20° | 25° | 30° | 35° | 20–30°C |
| *Brassica juncea* | Red light | 90 | 48 | 18 | 2 | — | 80 |
|                   | Dark | 53 | 20 | 6 | 0 | — | 34 |
| *Lepidium virginicum* | Red light | 21 | 32 | 33 | 0 | 0 | 29 |
|                       | Dark | 0 | 0 | 0 | 0 | 0 | 1 |
| *Nicotiana tabacum* | Red light | 94 | 96 | 94 | 84 | 8 | 97 |
|                     | Dark | 2 | 5 | 6 | 0 | 0 | 97 |
| *Zygophyllum dumosum* | White light | 77 | 84 | 72 | 12 | — | — |
|                       | Dark | 80 | 82 | 82 | 16 | — | — |
| *Calligonum comosum* | White light | 0 | 4 | 8 | 0 | — | — |
|                      | Dark | 0 | 64 | 80 | 12 | — | — |
| *Juncus maritimus* | White light | 62 | 74 | — | 70 | — | — |
|                    | Dark | 0 | 0 | — | 0 | — | — |

characterized by exposure to a low temperature followed by a high one. In the case of the *Paeonia* seeds the reverse requirement has been shown.

In *Convallaria majalis* full development of the seedling is extremely complicated. Root protrusion occurs at 25°C, but is greatly promoted if the seeds are first exposed to a low temperature, followed by a raised temperature to permit root growth. The first leaf enters dormancy after it has broken through the cotyledonous sheath and requires at this stage low temperature treatment at 5°C for its further development. Treatment at an earlier stage is ineffective. Further growth of the first leaf proceeds at normal temperature. The second leaf apparently also requires low temperature treatment for its development. Under normal conditions, in the natural habitat, such a cycle is not attained in less than 9 months and may take up to 14 months (Barton and Schroeder, 1942).

The literature on germination is full of examples where the temperature requirement is altered by exposure of the seeds to high, low or alternating temperatures. It seems clear that the changes occurring are extremely complex and affect primarily the rate of germination at the various temperatures. We have previously defined the optimal temperature of germination as involving both a rate factor and the actual germination percentage attained. Treatments with high, low or alternating temperatures can apparently widen the temperature range in which germination occurs; whether the optimal temperature is changed is not clear. This is illustrated by the behaviour of a number of seeds. *Betula lenta* seeds normally germinate only around 30°C. If the seeds are exposed to low temperature during stratification the temperature at which germination still occurs is reduced to 0°C. *Festuca* seeds will not germinate at 30°C when freshly harvested, but storage at 20–30°C for 1 year will cause them to germinate at 30°C. However, storage at low temperature does not have this effect. The seeds still fail to germinate at 30°C. Another instance is provided by annual *Delphinium* seeds, which will germinate at temperatures up to 30°C, following stratification at 5–10°C or even 15°C for 2 months, but not without this treatment.

### 3. *Light Requirements*

The fact that light can act as a dormancy-breaking agent has already been mentioned. The question of the mechanism of light stimulation of germination has also been discussed in Chapter 3. In addition to those seeds which require light for their germination, there are many species whose germination is inhibited by light (see Tables 4.4 and 4.6 as well as Table 3.8).

Light does not only affect the absolute germination percentage but also the rate of germination. Thus in *Agrostis* seeds the final germination percentage is directly related to the light intensity. However, at high light intensities it takes much longer to reach maximal germination than at low ones (Leggatt, 1946).

In all these cases, no matter whether light inhibits or promotes, there is a complex interaction between light and other external conditions as well as with the age of the seeds. For example, seeds of *Salvia pratensis, S. sylvestris, Epilobium angustifolium, Echium vulgare* and others lose their light sensitivity very shortly after harvest. In *Salvia verticillata, Epilobium parviflorum* and *Rumex acetosella* sensitivity is retained at least for a year. In other cases sensitivity is retained for much longer periods (Niethammer, 1927).

An instance of the complexity of the situation is provided by the case of lettuce seeds, variety Grand Rapids, which are sensitive to light. These seeds, immediately after harvest, hardly germinate at all in the dark at 26°C. After a certain period of storage they germinate in the dark at 18°C, but require a light stimulus for germination at 26°C. Peeling the seeds, including removal of the endosperm, or merely puncturing the endosperm, which encloses the embryo, is sufficient to abolish their light requirement at 26°C. This light requirement slowly decreases as the length of the storage period increases and eventually, after several years of storage, germination in the dark at 26°C reaches 60–80 per cent. It is possible to induce a light requirement in lettuce seeds which do not normally show it by treating the seeds with coumarin or by treatment with high temperatures or solutions of high osmotic pressure.

A more complicated interaction between light and temperature is shown by lettuce seeds, variety Grand Rapids. Seeds that gave about 30 per cent germination at 26°C, if kept for 2 days at 30°C and then returned to 26°C, only germinate about 1–4 per cent. Irradiation of such seeds with red light, following the treatment at 30°C, raises the germination percentage to 12–17 per cent, if the seeds are again returned to 26°C. Controls not given treatment at 30°C will germinate 95 per cent after this irradiation.

An induction of a light requirement was demonstrated by Toole (1959) for lettuce seeds, variety Great Lakes. These seeds germinate 95 per cent at 20°C in the dark. If kept for 4 days at 35°C they will only germinate 11 per cent in the dark at 20°C. However, irradiation with red light before returning them to 20°C restored germination to 95 per cent.

This interaction of light response of seeds with temperature was also clearly shown by Toole (1959) for seeds of *Lepidium virginicum*. Such seeds will not germinate in the dark if kept at 15°C or 25°C, or if transferred from 15 to 25°C or vice versa. If illuminated with red light, 2 days after sowing they will germinate to about 30–40 per cent, if kept throughout at either 15 or 25°C or on transfer from 25 to 15°C. However, the reverse transfer, from 15 to 25°C, when the light is given after the first 2 days, causes full germination. From this work and also from the examples quoted in Chapter 3, it is quite clear that the photochemical reaction in germination, involving phytochrome, is very closely linked to other, chemical, reactions, which will determine whether and to what extent there will be a response to illumination.

In many seeds the light effect is related to the presence of the seed coat. In lettuce seeds removal of the endosperm removes the light requirement. As will be seen from Table 4.4, the inhibition caused by light in the case of *Cucumis melo* is reduced by removal of the seed coat. In *Agropyron smithii*, removal of the seed coat does not abolish the inhibitory effect of light on germination. An unusual example is provided by *Oenothera biennis*. The intact seeds of this plant do not germinate in light or dark. However, if the seed coat is pierced, the seeds will germinate in light. Germination in the dark is only slightly promoted.

An entirely different light response is that of seeds whose germination is affected not by short illumination but by alternations between dark and light, i.e. by photoperiod. Photoperiodic responses of seeds were first shown by Isikawa (1954 and 1955). He suggests that both light-stimulated and light-inhibited seeds may be considered to show short day behaviour. Thus, in the light-requiring seeds of *Patrinia* and *Epilobium*, which germinate under illumination of from 2 to 21 hours, germination is promoted by interruption of the light period by darkness. Light-inhibited seeds such as *Nigella damascena* and *Silene armeria* germinate if illuminated for 1 minute or 3 hours daily respectively with low light intensities. The only instances quoted by Isikawa indicating a long day requirement are seeds of *Leptandra* and *Spiraea*.

A true long-day requirement for germination appears to exist in *Begonia* seeds. The seeds germinate if given at least three cycles of long days, the critical day length being 8 hours. A break in the dark period, if illumination is less than 8 hours, is also effective in causing germination. Furthermore, the response to light is increased in the presence of gibberellin, the critical day length being greatly shortened (Nagao *et al.*, 1959).

Other more complex examples are provided by the work on *Betula* (Black and Wareing, 1955), on *Tsuga canadensis* (Stearns and Olson, 1958) and on *Escholtzia argyi* (Isikawa and Ishikawa, 1960). *Betula* seeds will germinate at 15°C only under long day conditions, eight cycles being required to induce germination. However at 20–25°C germination will occur following a single exposure to light for 8–12 hours. *Escholtzia* seeds do not germinate either in the light or the dark at constant temperatures. The seeds will, however, germinate if the light treatment is combined with a thermoperiodic treatment. This is illustrated by Table 4.7. From this it appears that highest germination is obtained if a 6-hour photoperiod at 25°C is followed by an 18-hour thermoperiod at 5°C. The results shown in Table 4.7 are difficult to analyse as both photo- and thermoperiod were varied simultaneously. It is never clear to what change increased or decreased germination must be ascribed.

*Citrullus colocynthis* seeds provide another instance of photoperiodic response in germination. Seeds of this plant germinate in the dark at 20°C. However, exposure to long days, i.e. 12 hours of light per day, inhibits their germination. Short day treatment, 8 hours of illumination, does not inhibit. These phenomena thus show a marked similarity to photoperiodism in flowering. Some kind of induction phenomena have been observed in *Betula* at 15°C. What seems to be a clear case of induction exists in

Table 4.7—Percentage of Germination of Seeds of *Escholtzia*
with Daily Photoperiodic and Thermoperiodic Treatments
(After Isikawa and Ishikawa, 1960)
Each temperature and light treatment was repeated eight
times; the initial treatment was done in light

| Temperature and period of light exposure | | Temperature and period of darkness | | Germination (%) |
|---|---|---|---|---|
| 15°, | 6 hr | 5°, | 18 hr | 60 |
| 15°, | 18 hr | 5°, | 6 hr | 24 |
| 25°, | 6 hr | 5°, | 18 hr | 80 |
| 25°, | 18 hr | 5°, | 6 hr | 25 |
| 25°, | 6 hr | 15°, | 18 hr | 40 |
| 25°, | 18 hr | 15°, | 6 hr | 10 |
| 35°, | 6 hr | 5°, | 18 hr | 63 |
| 35°, | 18 hr | 5°, | 6 hr | 28 |
| 35°, | 6 hr | 15°, | 18 hr | 66 |
| 35°, | 18 hr | 15°, | 6 hr | 0 |

*Begonia* seeds (Nagao *et al.*, 1959). In these, germination occurs if the seeds are given six cycles of a photoperiod of 9 or more hours, followed by 7 days of darkness.

In *Eragrostis* seeds a dark period is required following imbibition in order that light becomes fully effective. Interruption of the dark period reduces its promotive effect. Short light periods, interrupted by dark, have been shown to be cumulative in their effect, for example in *Hypericum*. In *Atriplex dimorphostegia*, short periods of illumination promote while long periods of irradiation inhibit germination. Some of these effects are discussed by Koller (1972). In all cases, these photoperiodic effects of light appear to be related to the phytochrome system. It should be recalled that phytochrome interconversion can occur even in dry seeds, and that full imbibition is not required for photoconversion of phytochrome. Furthermore, repeated cycling between $P_R$ and $P_{FR}$ can eventually lead to the formation of inactive phytochrome. The details of the photoperiodic response of germination are still unclear, but in principle the rules relating to photoperiodic responses in general seem to apply.

## 4. *Germination Inhibitors*

A very large number of substances can inhibit germination. All those compounds which are toxic generally to living organism will also, at toxic concentrations, prevent germination, simply by killing the seed. Far more interesting is the action of those substances which prevent germination without affecting the seeds irreversibly. The simplest case, and one perhaps most frequently met with in nature, is that of osmotic inhibition. It is possible to prevent the germination of seeds simply by placing them in a solution of high osmotic pressure. When the seeds are removed from such an environment and placed in water, they can then germinate. Such a situation appears to exist in fruits and in the case of seeds of plants from saline habitats. The substances responsible for the high osmotic pressure may be sugars, inorganic salts such as sodium chloride, or other substances. The threshold of osmotic pressure at which germination is prevented, differs widely for different seeds. A few examples of these

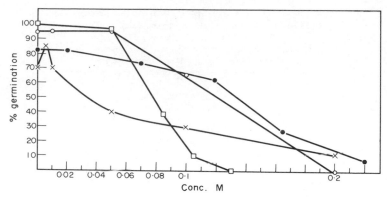

Fig. 4.3. The effect of sodium chloride concentration on the germination of various seeds.
(Compiled from data of Uhvits, 1946; Lerner *et al.*, 1959; and Poljakoff-Mayber,
unpublished)

×——× *Atriplex halimus*                    ●——● Alfalfa var.
                                                    Arizona Chilean
○——○ Tomato, var. Marmand          □——□ Lettuce
Germination percentages determined after 8 days for tomatoes, 4 days for *Atriplex* and 2 days
for lettuce and alfalfa.

differences are shown in Fig. 4.3. In the laboratory, a convenient compound for obtaining osmotic inhibition is mannitol. The use of mannitol frequently gives results somewhat different from those obtained with sodium chloride. This is clearly shown in Fig. 4.4 and may be ascribed to the ionic toxicity of the salts. It appears that the germination percentage obtained is not strictly a function of the water uptake by the seeds. As long as a certain minimal amount of water is taken up, germination can occur, provided that no toxic effects of the osmotic medium are present.

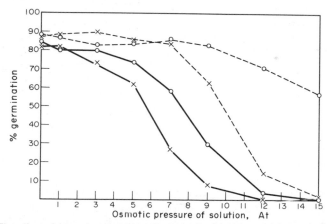

Fig. 4.4.  The effect of osmotic pressure and ionic toxicity on the germination of alfalfa seeds.
(From data of Uhvits, 1946)
○——○ Germinated in Mannitol for 2 days
○-----○ Germinated in Mannitol for 10 days
×——× Germinated in NaCl for 2 days
×-----× Germinated in NaCl for 10 days

Another type of inhibition is that caused by substances which are known to interfere with certain metabolic pathways. As germination is closely associated with active metabolism, all compounds which interfere with normal metabolism are likely to inhibit germination. An example of this type of interference is provided by various respiratory inhibitors. Compounds such as cyanide, dinitrophenol, azide, fluoride, hydroxylamine and others, all inhibit germination at concentrations similar to, but not always identical with, those inhibiting the metabolic processes. In some cases dinitrophenol and cyanide have a dormancy breaking effect. Cyanide is actively metabolized by some seeds if applied in low concentrations (Taylorson and Hendricks, 1973). These substances probably act on germination as a result of their effect on metabolism. The action of such compounds is not necessarily that of induction of dormancy. It is, however, interesting to note that Mancinelli (1958) claims that ethionine and hydroxylamine appear to increase the light requirement of lettuce seeds, or at any rate that light is able partially to reverse the inhibition caused by these substances. While the light reversal occurred both at 22°C and 26°C for ethionine, in the case of hydroxylamine the effect seemed more marked at 26°C.

Herbicides of various kinds inhibit germination to a greater or lesser extent. Many of the commonly used substances, such as 2,4-D, affect germination at comparatively low concentrations (Table 4.8). The more effective of such herbicides can and have

Table 4.8—Effect of 2,4-D and Coumarin on Germination of Various Seeds. The results are given as the concentration of the compound causing 50 per cent inhibition of germination (Compiled from data of Audus and Quastel, 1947; Mayer and Evenari, 1953; Isikawa, 1955; Libbert, 1957; Misra and Patnaik, 1959)

| Seed | Molar concentration | |
| | 2,4-D | Coumarin |
| --- | --- | --- |
| Cress | | $0·75 \times 10^{-4}$ |
| Radish | | $0·69 \times 10^{-4}$ |
| Mustard | | $> 1·3 \times 10^{-4}$ |
| Carrot | | $0·35 \times 10^{-4}$ |
| Beetroot | $5·0 \times 10^{-5}$ | $> 1·3 \times 10^{-4}$ |
| *Taraxacum* | | $0·41 \times 10^{-4}$ |
| Cabbage | | $0·27 \times 10^{-4}$ |
| Onion | | $> 1·3 \times 10^{-4}$ |
| Lettuce, var. Grand Rapids | $3·0 \times 10^{-5}$ | $1·5 \times 10^{-4}$ |
| Lettuce, var. Progress | $6·0 \times 10^{-5}$ | $5·0 \times 10^{-4}$ |
| *Setaria* | | $5·0 \times 10^{-4}$ |
| *Linum* | | $3·3 \times 10^{-5}$ |
| Rice | | $6·6 \times 10^{-3}$ |
| Wheat | $1·1 \times 10^{-3}$ | $1·5-3·0 \times 10^{-3}$ |

been used in order to prevent the germination of weed seeds in agricultural crops. No selective compound that will satisfactorily distinguish between crop and weed seeds appears to have been developed yet. A very frequent use of herbicides is as pre-emergence weed killers. In these cases the herbicide is applied in order to kill the seedling immediately after germination and before the main crop germinated. In these cases the herbicides are not in fact acting as germination inhibitors. Most growth retardants, such as Cycocel, Phosphon D and Amo 1618, if applied in sufficiently high concentrations also inhibit germination.

Phenolic compounds of various kinds inhibit germination. Some of the selective herbicides, such as the substituted phenols and cresols, inhibit germination due to their general phytotoxicity. In addition to these compounds, many other phenolic substances also inhibit germination. A list of a few of these and the concentration at which they inhibit the germination of lettuce seeds, is given in Table 4.9. Many other

Table 4.9—The Inhibition of Germination of Lettuce Seeds, Progress Variety, by a Number of Phenolic Compounds (Maycr and Evenari, 1952 and 1953)

| Compounds | Concentration causing 50% inhibition |
|---|---|
| Catechol | $10^{-2}$ M |
| Resorcinol | $5 \times 10^{-3}$ M |
| Salicylic Acid | $1 \cdot 5 \times 10^{-3}$ M |
| Gallic Acid | $5 \times 10^{-3}$ M |
| Ferulic Acid | $5 \times 10^{-3}$ M |
| Caffeic Acid | $> 10^{-2}$ M |
| Coumaric Acid | $5 \times 10^{-3}$ M |
| Pyrogallol | $10^{-2}$ M |

phenolic substances at similar concentration also affect the germination of a variety of seeds. Because of the widespread occurrence and distribution of phenolic compounds in plants and fruits, it has been suggested that these substances might act as natural germination inhibitors (van Sumere, 1960). This problem will be returned to later.

All compounds mentioned so far are inhibitory but cannot be regarded as being dormancy-inducing. The term dormancy-induction is used to imply that the seed can again be made to germinate by one or other of the treatments already mentioned earlier, which break natural dormancy. Coumarin has been widely used as a germination inhibitor in the laboratory. As already mentioned, coumarin can induce light sensitivity in varieties of lettuce seeds not requiring light for germination, as first shown by Nutile (1945).

Coumarin and its derivatives are of fairly widespread occurrence in nature. Coumarin itself (Fig. 4.5) is characterized by an aromatic ring and an unsaturated lactone structure. The effect of changes in structure of the coumarin molecule on its germination inhibiting activity has been investigated in some detail both for wheat and lettuce seeds. This work showed that there was no single essential group in the

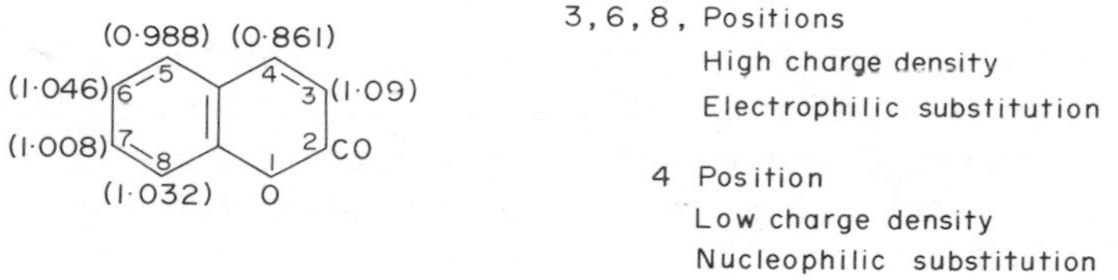

Fig. 4.5. Structure of coumarin with ring numbering and showing charge density.

coumarin molecule which was the cause of its inhibiting action. Reduction of the unsaturated lactone ring or substitution by most substituents such as hydroxyl, methyl, nitro, chloro and others in the ring system, all reduced the inhibitory activity of coumarin. A certain variation in the response of the two species tested were observed, which, however, did not materially alter the general picture (Mayer and Evenari, 1952). Opening of the lactone rings also reduced the inhibitory action of coumarin considerably.

The inhibitory action of coumarin has been studied on a wide variety of seeds and it has usually been found to inhibit germination. A few isolated instances of stimulation of germination by coumarin at very low concentrations are, however, known. The inhibitory concentration is different for different species and even differs in different varieties of the same species (Table 4.8). From these figures it will be seen that coumarin is extremely active as a germination inhibitor on a very wide variety of plant seeds. Because of its widespread distribution in plants and due to its strong inhibitory action, coumarin is considered to be one of the substances which may function as a natural germination inhibitor. This has been frequently proposed, but the occurrence of coumarin itself in seeds at inhibitory concentrations has been proved only in one instance, in the case of *Trigonella arabica* (Lerner *et al.*, 1959). However, coumarin derivatives such as the glycosides of the lactone, or substituted coumarins have been shown to occur in many fruits. For this reason the supposition that substances of this nature do function as germination inhibitors seems reasonable. Coumarin is actively metabolized in germinating seeds.

The most important other germination inhibitor is undoubtedly abscisic acid (Fig. 4.1). The presence of abscisic acid has been reported in very many seeds and fruits (Addicott and Lyon, 1969; Milborrow, 1974). Exogenously applied ABA prevents the germination of many species of seeds, as long as the ABA is present. The inhibition gradually disappears on removal of the seeds from the solution of ABA. The effect of ABA can also be removed more rapidly, by washing the seeds. The concentrations of ABA required to bring about its inhibitory effect are between 5–100 ppm. The effect of ABA may be regarded as that of a dormancy inducing compound. It may be acting by preventing embryo growth, for example in *Chenopodium album* (Karssen, 1968). The effect of abscisic acid on germination is sometimes reversed by gibberellic acid and in some cases also by cytokinins. The sensitivity of seeds to abscisic acid seems to be decreased in the light. ABA appears to be an important natural growth and germination inhibitor which interacts with the growth regulators in controlling seed dormancy and germination. However, ABA is clearly not the only natural growth inhibitor. For example, in tomato juice, the level of ABA present cannot account for the inhibitory activity of the juice. ABA did, however, increase osmotic inhibition (Dorffling, 1970).

Auxins in high concentrations generally inhibit germination. In some cases gibberellin has also been shown to inhibit germination (Fujii *et al.*, 1960).

Wiesner first suggested at the end of the nineteenth century that seeds of *Viscum* contain a germination inhibitor which prevents their germination within the fruit. This idea was further developed by other workers including Molisch. It appears that Oppenheimer (1922) was among the first to study this problem experimentally. He tested among other things the juice of tomatoes, to determine whether they contain a germination inhibitor and concluded that they did indeed contain such an inhibitory substance. This conclusion has since been disputed. Some authors claim that the inhibitory action is solely due to osmotic effects while others claim to have isolated the inhibiting substance. Thus Akkerman and Veldstra (1947) claim to have shown that caffeic and ferulic acids are the compounds responsible for the inhibition of germination of seeds within the tomato fruit. Despite this claim it must be pointed out that these authors failed to show that caffeic and ferulic acids occur in the tomato at concentrations sufficiently high to account for the inhibition of germination of the seeds in the mature fruit. It seems likely that the explanation of the failure of tomato seeds to germinate in the fruit can be ascribed to the combined action of osmotic inhibition together with a specific or non-specific germination inhibitor. Such interaction between the osmotic effect and the effect of an inhibitor has been shown to exist for example in the case of lettuce seeds (Lerner *et al.*, 1959). It seems very probable that this situation exists in many other cases also, as suggested by Evenari (1949), for example in grapes. The same author reviewed the older literature regarding germination inhibitors. Since this time a number of seeds and fruits have been investigated and some inhibiting substances isolated from them. The use of chromatographic techniques similar to those used for auxins has resulted in the detection of many as yet unknown substances, but a few have been identified. Thus coumarin and hydroxycinnamic acid and their derivatives, as well as vanillic acid, have been shown to occur in barley husks (van Sumere *et al.*, 1958). Both Varga (1957) and Ferenczy (1957) studied the inhibiting compounds occurring in lemons, strawberries and apricots. Coumarin and derivatives of cinnamic and benzoic acids were identified as the compounds responsible for inhibitory action. They suggested that the inhibitory activity is the result of the additive effect of a number of such compounds. The same type of compounds have also been identified in clusters of sugarbeet seeds and it is suggested that these substances here also act as germination inhibitors (Roubaix and Lazar, 1957; Massart, 1957).

A further germination inhibitor present in sugar beet fruits is cis-4-cyclohexene-1,2-dicarboximide (Mitchell and Tolbert, 1968). Just what the function of the complex mixture of compounds in the sugar beet fruit is and how they interact is still unclear.

Thus in most of the cases mentioned, it seems reasonable to accept the view that some substance or substances occur which inhibit germination. But, as already stated, the evidence regarding the chemical nature of these compounds is extremely vague, nor is it certain whether specific or non-specific compounds are involved. Evenari in his review attempts to classify the naturally occurring inhibitors found in seeds and fruits. He mentions cyanide-releasing complexes especially in Rosaceous seeds, ammonia-releasing substances, mustard oils, mainly in Cruciferae, various organic acids, unsaturated lactones, especially coumarin, parasorbic acid and protoanemonin, aldehydes, essential oils, alkaloids and phenols. From a glance at this list it can be seen that these compounds are not in any way restricted in occurrence to seeds and fruits. On the contrary substances of this general type occur in leaves, roots and other parts of plants as well as in the fruits and seeds. For this reason it is easy to understand why germination inhibitory substances have been isolated from a variety of plant tissues. Here again it is not clear what, if any, the biological importance of these substances is. This aspect of the problem will be dealt with in Chapter 7.

## III. Germination Stimulators

Various chemical substances can completely or partially substitute for light in breaking dormancy. These substances are very varied in chemical nature. Some of them are simple compounds, such as potassium nitrate and thiourea, while others are complicated molecules such as the gibberellins and cytokinins. The structure of a few of these dormancy-breaking substances is shown in Fig. 4.6.

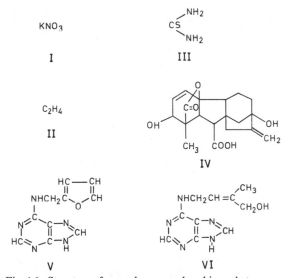

Fig. 4.6.  Structure of some dormancy-breaking substances.
  (I)   Potassium nitrate
  (II)  Ethylene
  (III) Thiourea
  (IV)  Gibberellin A3
  (V)   Kinetin
  (VI)  Zeatin

The effect of potassium nitrate on germination was discovered when it was noted that Knop's solutions promoted the germination of some seeds. Further experiments showed that potassium nitrate was responsible for this stimulation. Potassium nitrate promotes the germination of a number of seeds in the dark, e.g. *Lepidium virginicum, Eragrostis curvula, Polypogon monspelliensis*, various species of *Agrostis* and *Sorghum halepense*. The effect on some other species is shown in Table 4.10. The stimulation obtained by potassium nitrate is dependent on its concentration as shown in Fig. 4.7. As with light, potassium nitrate stimulation shows interaction with

Table 4.10—Effect of KNO₃ on Germination of Various Seeds
(After Hesse, 1924)
Concentration of KNO₃, 0·01 M

| Seed | Temp. | Incubation time in days | % germination | |
| --- | --- | --- | --- | --- |
| | | | water | KNO₃ |
| *Veronica longifolia* | 16–20° | 20 | 3·5 | 45 |
| *Hypericum perforatum* | 17–22° | 20 | 28·0 | 57 |
| *H. hirsutum* | 15–20° | 23 | 18·0 | 27 |
| *H. hirsutum* | 15–20° | 28 | 20·0 | 43 |
| *Epilobium hirsutum* | 16–20° | 14 | 23·0 | 40 |
| *E. montanum* | 16–20° | 14 | 6·5 | 89 |

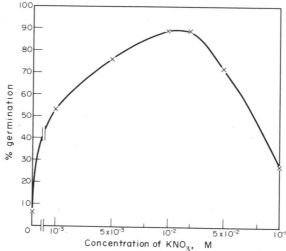

Fig. 4.7. Relation between germination percentage of seeds of *Epilobium montanum* and the potassium nitrate concentration (Germination in the dark for 14 days at 16°–20°). (After Hesse, 1924)

temperature. The germination of *Eragrostis curvula* is stimulated between 15 and 30°C in the dark by 0·2 per cent potassium nitrate. At higher temperatures or alternating temperatures there was no effect. In contrast the germination of *Polypogon* is promoted by potassium nitrate only at alternating temperatures (Toole, 1938).

The dark germination of many seeds is stimulated by thiourea. However, thiourea must be applied in high concentration in order to promote germination. Thus for lettuce seeds the effective concentration is of the order of $10^{-2}$ to $10^{-3}$M. These concentrations apply if the seeds are actually germinated in the solution of thiourea. A frequently adopted procedure is soaking the seeds in a concentrated solution of thiourea, 0·5 to 3 per cent, for a short time and then transferring them to water. Thus thiourea has been shown to stimulate the germination of seeds of *Cichorium*, *Gladiolus* (Shieri, 1941) and *Quercus*. In oak seed, as well as in other tree species such as *Larix* and *Picea*, thiourea substituted for the cold temperature treatment (Deubner, 1932; Johnson, 1946). Thiourea substitutes for the natural germination stimulator which develops in *Fraxinus* seeds during chilling (Villiers and Wareing, 1960). In some varieties of peach seeds, thiourea could substitute for after-ripening in promoting germination but subsequent seedling growth was abnormal (Tukey and Carlson, 1945). In lettuce and endive, thiourea abolished the inhibitory effect of high temperature; in lettuce thiourea also abolished the light requirements of the seeds. The effect of thiourea on the germination of lettuce was first described by Thompson and Kosar (1939) and has since been the subject of much study. Thiourea, in addition to stimulating germination, inhibits growth. Prolonged treatment of seeds with high concentrations of thiourea is therefore liable to cause an apparent inhibition of germination, because the emergence of the root is prevented. Thus it was found that germination of lettuce above 90 per cent was obtained, if the seeds were kept 24 hours in $5 \times 10^{-2}$M thiourea and then transferred to water. Water controls germinated only 40 per cent. Keeping the seeds longer in thiourea reduced the germination below 90 per cent. As in many other effects on germination, here also there exist interactions between the effect of thiourea and that of other factors affecting germination, such as light and temperature. Moreover, the effects observed are a function of the thiourea concentrations (Table 4.11). It is clear that at different

Table 4.11—The Effect of Thiourea on the Germination of Lettuce, Variety Grand Rapids, at Different Concentrations and Temperatures (Poljakoff-Mayber *et al.*, 1958)

|  | % germination | |
| --- | --- | --- |
| Thiourea | 20°C | 26°C |
| 0 | 26·8 | 5·8 |
| $1·0 \times 10^{-2}$ M | 67·4 | 21·4 |
| $2·5 \times 10^{-2}$ M | 76·5 | 16·1 |
| $5·0 \times 10^{-2}$ M | 56·3 | 12·4 |

temperatures the concentration of thiourea effective in causing germination is different. An interaction between light and thiourea was shown by Evenari *et al.* (1954). The effects of thiourea on the metabolism of seeds, which have been studied in some detail, will be discussed in Chapter 6. The action of derivatives of thiourea on the germination of lettuce seeds has also been studied. All modifications of the structure of thiourea, e.g. ethyl, methyl and phenyl substitution, converted it from a germination stimulator to a substance inhibiting germination (Mayer, 1956). Thiourea has been shown to be present in the seeds of at least one plant species, *Laburnum anagyroides* (Klein and Farkass, 1930). The effect on germination of a large number of other thiol compounds was studied by Reynolds (1974).

Ethylene chlorhydrin has been shown in a number of cases to have similar effects on germination as those of thiourea. However, it is usually less effective. Both substances were first used to break dormancy in buds and only at a later stage used for breaking dormancy in seeds. Ethylene has been shown to stimulate the germination of a number of seeds. Whether this stimulation is a dormancy breaking action is still uncertain. Ethylene may act by affecting other growth substances. Certainly ethylene interacts with other substances in its effect on germination. Ethylene can be applied to seeds as 2-chloroethane phosphonic acid (Ethrel), which liberates ethylene in the presence of water at pH above 4·0.

That other compounds of stimulating activity are of widespread occurrence in nature seems likely. Libbert and Lubke (1957) showed that the widely distributed scopoletin can promote germination in old seeds of *Sinapis alba* at very low concentrations. Ruge (1939) indicated that in *Helianthus annuus* seeds, there is a balance between a germination-stimulating and an inhibiting factor, without, however, identifying either of them. The best established cases of natural stimulators are those of the factors responsible for stimulating the germination of the seeds of the parasitic plants *Striga sp.* and *Orobanche sp.* Brown and his co-workers (1952) showed that seeds of both these plants germinate in the vicinity of the host only after the latter secretes a germination-stimulating substance. It appears that one of these stimulatory compounds may be xylulose. A number of additional substances seem to have been partially isolated from exudates of host plants of *Orobanche* and *Striga*. Unfortunately these substances have not yet been identified chemically (Edwards, 1972).

### IV.  Hormones in Germination

In a number of isolated cases hormones have been shown to promote germination. Although this effect has been known for some time, it is by no means clear-cut. The term hormone is here used in the widest sense of the word, including various growth substances and natural compounds having a regulatory function in the plant. Most of the dubious results have been obtained using auxin. Much clearer effects have been obtained by the external application of gibberellins or gibberellic acid, but the effect is by no means universal. Lona (1956) and Kahn *et al.* (1956) were among the first to show that gibberellic acid stimulates the germination of *Lactuca scariola* and *Lactuca sativa* in the dark. In these cases gibberellic acid substituted for light in promoting germination. Other cases of substitution of gibberellic acid for red light are known in the case of seeds of *Arabidopsis, Kalanchoe* and *Salsola volkensii.* However, in a number of seeds whose germination is either promoted or inhibited by light, e.g. *Juncus maritimus, Oryzopsis miliacea* and others, treatment with gibberellic acid was not effective in promoting germination (Leizorowitz, 1959). Recently the germination of a number of species of plants whose germination is not affected by light have been shown to be promoted by gibberellic acid, as shown in Table 4.12. The sensitivity of some of the seeds depended on the time of harvest. There was no relation between the place of the seeds in the systematic classification of plants and their response to gibberellic acid.

The effect of gibberellic acid seems to be similar to that of light in promoting germination (Table 4.13). But gibberellic acid is far more effective than red light in reversing the high temperature inhibition of germination in lettuce. Thus 100 ppm

Table 4.12—Effect of Gibberellic Acid on the Germination of Various
Seeds
(Compiled from data of Kallio and Piiroinen, 1959; Corns, 1960)

| | | % germination | |
| Plant | Treatment with G.A. | Treated | Water control |
| --- | --- | --- | --- |
| *Avena fatua* | 500 ppm | 57 | 26 |
| *Sinapis arvensis* | 500 ppm | 89 | 9 |
| *Thlapsi arvense* | 500 ppm | 90–100 | 0 |
| *Gentiana nivali* | 1000 ppm (24 hr) | 90 | 0 |
| *Bartschia alpina* | 1000 ppm (24 hr) | 90 | 5 |
| *Draba hirta* | 1000 ppm (24 hr) | 95 | 0 |

Table 4.13—Effect of Light and Gibberellic Acid
on the Germination of Lettuce at 26°C
(Evenari *et al.*, 1958)

| | % germination | |
| Treatment | In water | In gibberellic acid $2·9 \times 10^{-5}$ M |
| --- | --- | --- |
| Dark | 12·0 | 39·0 |
| Red | 44·0 | 66·5 |
| Far-red | 5·0 | 25·0 |
| Red and far-red | 11·0 | 35·0 |

gibberellic acid stimulated the germination of lettuce seeds at 30°C from 2 per cent in the water to 33 per cent. In contrast red light only stimulated the germination from 2 to 4 per cent. When both gibberellic acid and light were given, the germination percentage obtained was 50. Gibberellic acid can also reverse the inhibition of germination caused by high osmotic pressure. Kahn (1960) showed that lettuce seeds germinating to 82 per cent in the dark, in water, gave only 22 per cent germination in 0·15M mannitol. However, addition of 35 ppm gibberellic acid to the mannitol resulted in a germination percentage of 61, showing a reversal of osmotic inhibition. Similar reversals of osmotic inhibition were obtained by using red light, from 36 per cent in 0·18M mannitol to 88 per cent in mannitol plus red light. Originally it was assumed that red light and gibberellic acid act in an entirely similar fashion in promoting germination of lettuce seeds. However, more detailed investigation along the lines mentioned, studying the interaction of gibberellic acid with far-red light and with high temperatures, showed that this was not the case (Evenari *et al.*, 1958). Thus, it is not always possible effectively to reverse the gibberellic acid induced stimulation by the use of far-red light, indicating that part of the gibberellic acid induced germination differs from that induced by red light. These experiments have resulted in the general conclusion that gibberellic acid and red light act only partially in the same way and that their mode of action is not identical (Ikuma and Thimann, 1960; Fujii *et al.*, 1960). That gibberellic acid might be of importance in determining the germination of seeds in nature is indicated by the isolation from *Phaseolus*, lettuce and many other seeds of gibberellinlike compounds (Phinney and West, 1960). There is some evidence to indicate that the amount of gibberellinlike substances changes during different stages of germination and during ripening or after-ripening of the seeds.

Some of the known 29 gibberellins occur in bound form. During fruit and seed ripening some gibberellins are converted from the free to the bound form and this process may be reversed during germination (Lang, 1970).

The origin of the gibberellin in the seed is different in different seeds. In barley gibberellic acid-like substances are formed in the scutellum and then move out of the embryo into the aleurone layer and the endosperm. However, some additional factor is required, emanating from the axis, in order to initiate scutellar synthesis of gibberellins (Radley, 1968). In peas, on the other hand, the gibberellins are apparently released from a bound form in the cotyledons and move to the embryonic axis (Barendse *et al.*, 1968). Much remains to be discovered about the action of gibberellins in dormancy breaking and its interaction with environmental and other factors. The known effects of the gibberellins on metabolism during germination will be discussed in Chapter 6.

Another plant hormone, cytokinin, also affects germination of seeds. Miller (1958) showed that kinetin promotes the germination of seeds. Later the natural cytokinin, zeatin was isolated and characterized (see Fig. 4.6) (Letham *et al.*, 1964). The natural cytokinin acts in the same way as kinetin. In addition to zeatin its ribotide and riboside have been described. In germination stimulation, these derivatives are less active than zeatin itself (Van Staden, 1973). The cytokinins are actively metabolized in germinating seeds. A large number of derivatives of kinetin, where the furfuryl group is replaced by other groupings, also stimulate germination, for example benzyladenine. While originally it was thought that kinetin substitutes for red light in germination, it was later shown that in fact the seeds were sensitized by kinetin so that a smaller dose of light would induce their germination. Weiss (1960) has shown that while in water lettuce seeds required 3,600 f.c. sec. light to promote their germination, in the presence of kinetin 720 f.c. sec. light were sufficient to bring about the same effect. The seeds need not be kept continuously in a solution of kinetin. As little as a few hours in the solution, at a suitable period, is sufficient to cause the sensitization. That kinetin does not substitute for light is also indicated by the fact that light-inhibited seeds such as *Oryzopsis miliacea* are stimulated by kinetin, while other light sensitive seeds such as *Amaranthus* are not affected by kinetin at all.

It seems likely that the requirement of light for the full expression of the stimulatory effect kinetin depends on the development of some factor in the seed immediately upon illumination. This may be some direct or indirect result of the phytochrome reaction (Reynolds and Thompson, 1973). In addition to their interaction with light, cytokinins also interact with other exogenously applied compounds, such as ABA and GA and also with temperature.

The effect of hormones of the indolylacetic acid (IAA) type on germination has long been in dispute. Numerous workers have investigated the effect of IAA and similar substances on the germination of a variety of seeds, and have obtained conflicting results, stimulation or inhibition being obtained, depending on the concentration of IAA and the type of seed used. However, the most general effect is an absence of response of the seeds to physiological concentrations of IAA. Soeding and Wagner (1955) attempted to relate divergent observations to the state of dormancy of the seeds. However, they failed experimentally to prove such a relation for seeds of *Poa*. A relation between depth of dormancy and response to IAA was, however, established for lettuce seeds. Even here the observed absolute effects were usually very small. In seeds where germination in the dark was of the order of 4–10 per cent, IAA at a concentration of $10^{-7}M$ raised the germination in the dark to 20–30 per cent, while on seeds germinating 60–80 per cent in the dark, IAA had no effect whatever. In other experiments the response of the seeds to IAA was found to be temperature-dependent. At 20° $10^{-7}M$ IAA raised the germination of lettuce seeds from 27 to 47 per cent, while at 26°C the germination percentage was 8, both in water and in IAA (Poljakoff-Mayber, 1958).

Thus from these results some relationship between response to IAA, dormancy and germination conditions emerges, but the effects are in no way clear cut. It is not possible to relate these results to internal IAA concentrations, as dry lettuce seeds apparently do not contain IAA. In general, therefore, it must be concluded that IAA can, under very special conditions, stimulate germination, but normally it has little or no effect.

From the above discussion on inhibitors and stimulators it may be concluded that the germination of seeds is controlled by a variety of external and internal factors. These factors include, in addition to simple environmental conditions, also the presence of external and internal germination-inhibiting and stimulating substances. One of the controlling factors is the balance between stimulatory and inhibitory concentration of the compounds at their site of action. These cannot be deduced from experiments with exogenously applied substances. There is always uncertainty of their permeability into the seed and the extent of their metabolism.

# Bibliography

Addicott, F. and Lyon, J. L. (1969) *Ann. Rev. Plant Physiol.* **20**, 139.
Akkerman, A. M. and Veldstra, H. (1947) *Rec. Trav. Chim.* **66**, 411.
Audus, L. J. and Quastel, J. H. (1947) *Nature, Lond.* **159**, 230.
Axentjev, B. N. (1930) *B.B.C.* **46**, 119.
Ballard, L. A. T. (1958) *Austr. J. Biol. Sci.* **11**, 246.
Barendse, C. W. M., Kende, H. and Lang, A. (1968) *Plant Physiol.* **43**, 815.
Barton, L. V. (1934) *Contr. Boyce Thompson Inst.* **6**, 69.
Barton, L. V. and Crocker, W. (1948) *Twenty Years of Seed Research*, Faber and Faber, London.
Barton, L. V. and Schroeder, E. M. (1942) *Contr. Boyce Thompson Inst.* **12**, 277.
Black, M. and Wareing, P. F. (1955) *Physiol. Plant.* **8**, 200.
Brown, R. (1940) *Ann. Bot. N.S.* **4**, 379.
Brown, R. (1946) *Nature, Lond.* **157**, 64.
Brown, R., Johnson, A. W., Robinson, E. and Ryler, G. (1952) *Biochem. J.* **50**, 596.
Brown, R., Greenwood, A. D., Johnson, A. W., Landsdown, A. R., Long, A. G. and Sunderland, N. (1952) *Biochem. J.* **52**, 571.
Capon, B. and Van Asdall, W. (1967) *Ecology*, **48**, 305.
Come, D. (1968) *Bull. Soc. Franc. Physiol. Veg.* **14**, 31.

Corns, W. G. (1960) *Canad. J. Plant. Sci.* **40**, 47.
Crocker, W. (1906) *Bot. Gaz.* **42**, 265.
Crocker, W. and Barton, L. V. (1953) *Physiology of Seeds*, Chronica Botanica.
Davis, W. E. (1930) *Amer. J. Bot.* **17**, 58.
Deubner, C. G. (1932) *J. Forestry*, **30**, 672.
Dorffling, K. (1970) *Planta* **93**, 243.
Eckerson, S. (1913) *Bot. Gaz.* **55**, 286.
Edwards, M. M. (1969) *J. Expt. Bot.* **20**, 876.
Edwards, W. G. H. (1972) In *Phytochemical Ecology*, p. 235 (ed. J. B. Harborne), Acad. Press, London.
Esashi, Y. and Leopold, A. C. (1968) *Plant Physiol.* **43**, 871.
Evenari, M. (1949) *Bot. Rev.* **15**, 143.
Evenari, M., Stein, G. and Neuman, G. (1954) *Proc. Ist. Int. Photobiological Congress*, p. 82, Amsterdam.
Evenari, M., Neuman, G., Blumenthal-Goldschmidt, S., Mayer, A. M. and Poljakoff-Mayber, A. (1958) *Bull. Res. Council Israel* **6D**, 65.
Ferenczy, I. (1957) *Acta Biol. Hung.* **8**, 31.
Flemion, F. (1933) *Contr. Boyce Thompson Inst.* **5**, 143.
Flemion, F. and Prober, P. L. (1960) *Contr. Boyce Thompson Inst.* **20**, 409.
Flemion, F. and De Silva, D. S. (1960) *Contr. Boyce Thompson Inst.* **20**, 365.
Frankland, B. and Wareing, P. F. (1966) *J. Expt. Bot.* **17**, 596.
Fujii, T., Isikawa, S., and Nakagawa, A. (1960) *Bot. Mag. Tokyo* **73**, 404.
Grant Lipp, A. E. and Ballard, L. A. T. (1959) *Austr. J. Agr. Res.* **10**, 495.
Gutterman, Y., Witztum, A. and Evenari, M. (1967) *Is. J. Bot.* **16**, 213.
Hamly, D. H. (1932) *Bot. Gaz.* **93**, 345.
Hesse, O. (1924) *Bot. Archiv.* **5**, 133.
Ikuma, H. and Thimann, K. V. (1960) *Plant Physiol.* **35**, 557.
Ikuma, H. and Thimann, K. V. (1963) *Plant and Cell Physiol.* **4**, 169.
Isikawa, S. (1954) *Bot. Mag. Tokyo* **67**, 51.
Isikawa, S. (1955) *Kumamoto J. Science Ser.* **B2**, 97.
Isikawa, S. and Ishikawa, T. (1960) *Plant and Cell Physiol.* **1**, 143.
Isikawa, S., Fujii, T. and Yokohama, Y. (1961) *Bot. Mag. Tokyo* **74**, 14.
Johnson, L. P. V. (1946) *Forestry Chronicle* **22**, 17.
Kahn, A., Goss, J. A. and Smith, D. E. (1956) *Plant Physiol.* **31**, suppl. 37.
Kahn, A. (1958) *Plant Physiol.* **33**, 115.
Kahn, A. (1960) *Plant Physiol.* **35**, 1.
Kahn, A. (1960) *Plant Physiol.* **35**, 333.
Kallio, P. and Piiroinen, P. (1959) *Nature, Lond.* **183**, 1930.
Karssen, C. M. (1968) *Acta Bot. Neerl.* **17**, 293.
Khudairi, A. K. (1956) *Physiol. Plant* **9**, 452.
Kidd, F. (1914) *Proc. Roy Soc.* **B87**, 609.
Klein, G. and Farkass, E. (1930) *Öst. Bot. Z.* **79**, 107.
Koller, D. (1954) Ph.D. Thesis Hebrew University, Jerusalem (in Hebrew).
Koller, D. (1972) In *Seed Biology*, vol. 2, p. 2 (ed. T. T. Kozlowski), Acad. Press, New York.
Lang, A. (1970) *Ann. Rev. Plant Phys.* **21**, 537.
Leggatt, C. W. (1946) *Can. J. Research* **C24**, 7.
Leizorowitz, R. (1959) M.Sc. Thesis, Jerusalem (in Hebrew).
Lerner, H. R., Mayer, A. M. and Evenari, M. (1959) *Physiol. Plant* **12**, 245.
Letham, D. S., Shannon, J. S. and McDonald, I. R. (1964) *Proc. Chem. Soc.*, p. 230.
Libbert, E. and Lubke, H. (1957) *Flora* **145**, 256.
Libbert, E. (1957) *Phyton* **9**, 81.
Lona, F. (1956) *Ateneo Parmense* **27**, 641.
Maier, W. (1933) *Jahr. Wiss. Bot.* **78**, 1.
Mancinelli, A. (1958) *Ann. di Botanica* **26**, 56.
Marbach, I. and Mayer, A. M. (1974) *Plant Physiol.* **54**, 817.
Marchaim, U., Birk, Y., Dovrat, A. and Berman, T. (1972) *J. Expt. Bot.* **23**, 302.
Mato, N. (1924) *Sitz. Acad. Wiss. Matt. Natur. K.L.* **133**, 625.
Mayer, A. M. and Evenari, M. (1952) *J. Expt. Bot.* **3**, 246.
Mayer, A. M. and Evenari, M. (1953) *J. Expt. Bot.* **4**, 257.
Mayer, A. M. (1956) *J. Expt. Bot.* **7**, 93.
Milborrow, B. V. (1974) *Ann. Rev. Plant Physiol.* **25**, 259.
Miller, C. O. (1958) *Plant. Physiol.* **33**, 115.

Misra, G. and Patnaik, S. N. (1959) *Nature, Lond.* **183**, 989.
Mitchell, E. D. and Tolbert, N. E. (1968) *Biochemistry,* **7**, 1019.
Nagao, M., Esashi, Y., Tanaka, T., Kumaigai, T. and Fukumoto, S. (1959) *Plant and Cell Phys.* **1**, 39.
Niethammer, A. (1927) *Biochem. Z.* **185**, 205.
Nikolaeva, M. G. (1969) *Physiology of deep dormancy in seeds.* Israel Program Sci. Trans. Jerusalem.
Nutile, G. E. (1945) *Plant Physiol.* **20**, 433.
Olney, H. O. and Pollock, B. M. (1960) *Plant Physiol.* **35**, 970.
Oppenheimer, H. (1922) *Sitz. Akad. Wiss. Wien Abt.* **1 131**, 279.
Phinney, B. O. and West, C. A. (1960) *Ann. Rev. Plant Physiol.* **11**, 411.
Poljakoff-Mayber, A. (1958) *Bull. Res. Council Israel.* **6D**, 78.
Poljakoff-Mayber, A., Mayer, A. M. and Zachs, S. (1958) *Ann. Bot. N.S.* **22**, 175.
Pollock, B. M. and Olney, H. O. (1959) *Plant Physiol.* **34**, 131.
Porter, N. G. and Wareing, P. F. (1974) *J. Expt. Bot.* **25**, 583.
Radley, M. (1968) Soc. Chem. Ind. (London) Monograph **31**, 53.
Reynolds, T. (1974). *J. Expt. Bot.* **25**, 375.
Reynolds, T. and Thompson, P. A. (1973) *Physiol. Plant.* **28**, 516.
Roubaix, J. de and Lazar, O. (1957) *La Sucrerie Belge* **5**, 185.
Ruge, U. (1939) *Z.f.Bot.* **33**, 529.
Shieri, H. B. (1941) *The Gladiolus* **16**, 100.
Shull, C. A. (1911) *Bot. Gaz.* **52**, 455.
Shull, C. A. (1914) *Bot. Gaz.* **57**, 64.
Soeding, H. and Wagner, M. (1955) *Planta* **45**, 557.
Spaeth, J. N. (1932) *Amer. J. Bot.* **19**, 835.
Stearns, F. and Olsen, J. (1958) *Amer. J. Bot.* **45**, 53.
Sumere, C. F. van, Hilderson, H. and Massart, L. (1958) *Naturwiss.* **45**, 292.
Sumere, C. F. van (1960) In *Phenolics in Plants in Health and Disease,* p. 25, Pergamon Press, Oxford.
Taylorsen, R. B. and Hendricks, S. D. (1973) *Plant Physiol.* **52**, 23.
Thompson, R. C. and Kosar, W. F. (1939) *Plant Physiol.* **14**, 561.
Thornton, N. C. (1935) *Contr. Boyce Thompson Inst.* **7**, 477.
Toole, E. H. (1959) In *Photoperiodism and related phenomena in plants and animals,* AAAS Publ. No. 55 (ed. Withrow).
Toole, E. H., Toole, V. K., Borthwick, H. A. and Hendricks, S. B. (1955) *Plant Physiol.* **30**, 473.
Toole, V. K. (1938) *Proc. Ass. Off. Seed Analyst. N. America,* p. 227, 30th Ann. Meeting.
Tukey, H. B. and Carlson, R. F. (1945) *Plant Physiol.* **20**, 505.
Uhvits, R. (1946) *Amer. J. Bot.* **33**, 278.
Van Staden, J. (1973) *Physiol. Plant.* **28**, 222.
Varga, M. (1957) *Acta Biol. Hung.* **7**, 39.
Varga, M. (1957) *Acta Biol. Szeged.* **3**, 213.
Varga, M. (1957) *Acta Biol. Szeged.* **3**, 225.
Varga, M. (1957) *Acta Biol. Szeged.* **3**, 233.
Villiers, T. A. and Wareing, P. F. (1960) *Nature, Lond.* **185**, 112.
Wareing, P. F. and Foda, H. A. (1957) *Physiol. Plant.* **10**, 266.
Weiss, J. (1960) *C.R. Acad. Sci. Paris* **251**, 125.
Wiesner, J. (1897) *Ber. Deut. Bot. Ges.* **15**, 503.
Wycherley, P. R. (1960) *J. Rubb. Res. Inst. Malaya* **16**, 99.

Chapter 5

# METABOLISM OF GERMINATING SEEDS

The dry seed is characterized by a remarkably low rate of metabolism. This is probably a direct result of the almost complete absence of water in the seed, whose water content is of the order of 5–10 per cent. Despite this almost complete absence of metabolism in the seed, it cannot be assumed that it lacks the potentiality for metabolism. When dry seeds are broken up by grinding in a suitable aqueous medium it is possible to show in the extract the presence of a considerable number of enzyme systems. Thus it can be concluded that the dry seed is a well equipped functional unit which can carry out a large number of biochemical reactions, provided that the initial hydration of the proteins and more specifically of the enzyme proteins, has taken place.

One of the most frequently used criteria for determining the rate of metabolism is the rate of respiration. Dry seeds show a very low rate of respiration, which begins to rise rapidly when the seeds are placed in water. Thus the first easily observable metabolic change in the seed, well before germination occurs, is the increase in the respiratory rate from values close to zero to appreciable rates. The second effect, which is very easily observed, is the breakdown of reserve materials in the seed. This reaction again is initiated by the hydration of the proteins. As long as the seeds are dry almost no changes whatever are observed in their chemical composition. As soon as the seed is hydrated very marked changes in composition in its various parts occur. The chemical changes which occur are complex in nature. They consist of three main types: the breakdown of certain materials in the seed, the transport of materials from one part of the seed to another and especially from the endosperm to the embryo, or from the cotyledons to the growing parts, and lastly the synthesis of new materials from the breakdown products formed. Of special importance is the initiation of protein synthesis during germination and its relation to nucleic acid metabolism. At the ultrastructural level many changes occur during germination and such changes must in some way be related to the overall metabolism of the germinating seed. The only substances normally taken up by seeds during germination are water and oxygen. In many instances substances are lost from the seed during initial stages of germination. The initial stages of germination are consequently accompanied by a net loss of dry weight due to the oxidation of substances on the one hand and leakage out of the seed on the other hand. Only when the seedling has formed, i.e. when a root is present which takes up minerals, and the cotyledons or first leaves are exposed to light and capable of photosynthesis, does an increase in dry weight begin.

In the following we will discuss the breakdown of reserve materials in seeds, respiration, protein metabolism and the general metabolism of seed components. The complex effect of growth substances, germination inhibitors and stimulators and their role in the control of the initiation of germination will be discussed in a separate chapter, Chapter 6.

## I. Changes in Storage Products During Germination

As already mentioned in Chapter 2, the chief types of seeds are those containing fat and those containing carbohydrates as a reserve material. A variety of other materials are also present in the dry seeds as already discussed in Chapter 2. Although some of these compounds are present in only small amounts, their metabolism is nevertheless of considerable importance during germination. A number of other compounds do not seem to be metabolized to any appreciable extent.

Changes in the composition of seeds during germination have been investigated in a number of species. Among the more detailed studies are those of Oota *et al.* (1953) on *Vigna sesquipedalis*, rice (Palmiano and Juliano, 1972), corn (Ingle *et al.*, 1964, 1965) and Douglas fir (Ching, 1966).

A very clear illustration of the changes in composition in different parts of a seed during germination is given in Figs. 5.1–5.8. These figures are taken from the detailed investigations of Oota *et al.* (1953) on the metabolism of germinating beans, *Vigna sesquipedalis*. This seed is characterized by large cotyledons and epigeal germination. The cotyledons show a loss of all compounds studied, while all the other organs show an increase in the various seed constituents. Particularly striking is the increase of materials in the rapidly growing hypocotyl, which ceases as the hypocotyl ceases to grow. The seed studied by Oota lends itself particularly to such detailed analysis as it is technically feasible to differentiate between hypocotyl, root, cotyledons, plumule and epicotyl.

As will be seen from Fig. 5.1 the cotyledons show a steady decrease in dry weight for

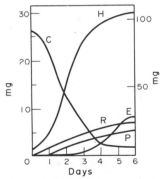

Fig. 5.1. Changes in dry weight of embryonic organs of *Vigna sesquipedalis* during germination.
Values per individual are plotted. P—a pair of plumules; E—epicotyl; H—hypocotyl; R—radicle; C—a pair of cotyledons. Scales on the right-hand ordinate for C; on the left-hand ordinate for P, E, H and R.
(From Oota *et al.*, 1953)

the first 4 days of germination. At the same time there is a similar increase in dry weight in the other parts of the seedling and especially the hypocotyl. The metabolism of the radicle appears to start almost immediately on hydration. When the hypocotyl ceases growing, after 3 or 4 days, a rapid rise in dry weight of the epicotyl is observed. All marked changes in the epicotyl are delayed till the onset of its growth. During the 6 days of measurement about 20 per cent dry weight is lost from the seed, presumably due to respiration. A study of Fig. 5.2 and Fig. 5.8 shows a very similar situation for

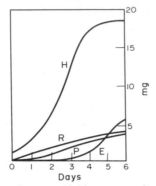

Fig. 5.2. Changes in content of water-soluble matter of embryonic organs of *Vigna sesquipedalis* during germination.
Values per individual are plotted.
(Symbols as Fig. 5.1)
(From Oota *et al.*, 1953)

total water soluble and insoluble material in the embryonic organs, but accumulation of insoluble material in the hypocotyl continues when that of soluble material has already ceased. Carbohydrates accumulate in the various parts of the embryo in a similar fashion, but as will be seen in Fig. 5.3, after 4 days soluble sugars again

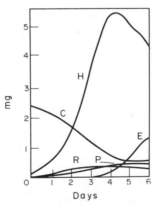

Fig. 5.3. Changes in content of soluble sugar, expressed as equivalents of glucose, of embryonic organs of *Vigna sesquipedalis* during germination.
Values per individual are plotted.
(Symbols as in Fig. 5.1)
(From Oota *et al.*, 1953)

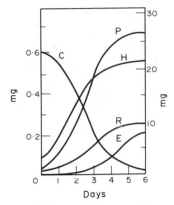

Fig. 5.4. Changes in content of insoluble polysaccharide (expressed as equivalents of glucose) of embryonic organs of *Vigna sesquipedalis* during germination.
Values per individual are plotted.
(Symbols as in Fig. 5.1)
Scales on the right-hand ordinate for C; on the left-hand ordinate for P, E, H and R.
(From Oota *et al.*, 1953)

decrease in the hypocotyl and begin to appear in the epicotyl. Insoluble polysac-charides, consisting primarily of starch and dextrin, however, do not show this decrease in the hypocotyl (Fig. 5.4). The pattern for soluble and protein nitrogen (Fig. 5.5 and Fig. 5.7) is again very similar, but the increase in protein nitrogen lags behind

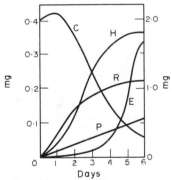

Fig. 5.5. Changes in content of soluble nitrogen of embryonic organs of *Vigna sesquipedalis* during germination.
Values per individual are plotted.
(Symbols as in Fig. 5.1)
Scales on right-hand ordinate for H; on left-hand ordinate for P, E, R and C.
(From Oota *et al.*, 1953)

the increase in soluble nitrogen. The decrease in soluble nitrogen of the cotyledons does not begin till the second day. In fact, soluble nitrogen shows a distinct peak after 2 days, presumably due to an unequal rate of protein breakdown of transport of the breakdown products from the cotyledons. Fig. 5.6 shows the changes in RNA (ribonucleic acid) in the different parts of the embryo. The cotyledons lose steadily but

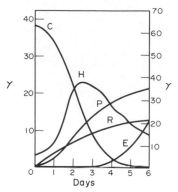

Fig. 5.6. Changes in content of pentose nucleic acid phosphorus of embryonic organs of
*Vigna sesquipedalis* during germination.
Values per individual are plotted.
(Symbols as in Fig. 5.1)
Scales on the right-hand ordinate for C; on the left-hand ordinate for P, E, H and R.
(From Oota *et al.*, 1953)

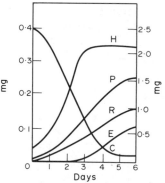

Fig. 5.7. Changes in content of protein nitrogen of embryonic organs of *Vigna sesquipedalis*
during germination.
Values per individual are plotted.
(Symbols as in Fig. 5.1)
Scales on the right-hand ordinate for C; on the left-hand ordinate for P, E, H and R.
(From Oota *et al.*, 1953)

the hypocotyl shows a peak in RNA between the second and third day and a
subsequent linear decrease. The peak appears before the cessation of hypocotyl
growth. The data brought above show a marked mobility of the various components of
the seed. The function of the cotyledons as storage organs, which empty out as other
parts of the embryo develop, is clearly demonstrated.

In corn, dry weight and total N of the whole seedling decrease during the first 120
hours of germination. The drop is primarily in the endosperm while total N and dry
weight increase in the axis. Insoluble protein shows similar changes. However, soluble
protein and total amino acid rise in the whole seedling and the axis, while soluble
protein in the endosperm shows a peak after 3 days. Nucleic acids and soluble

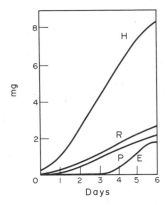

Fig. 5.8. Changes in content of water insoluble matter of embryonic organs of *Vigna sesquipedalis* during germination.
Values per individual are plotted.
(Symbols as in Fig. 5.1)
(From Oota *et al.*, 1953)

nucleotides rise both in the whole seedling and the axis during the same time period, but remain at a constant level in the endosperm and scutellum. A more detailed investigation showed that the overall RNA and DNA content rises for 4 days of germination (Ingle and Hageman, 1965). This rise occurs in the axis and the scutellum, but a decrease was noted in the endosperm. Soluble nucleotides increased in all parts of the seedling. Generally, changes in the scutellum of corn are small (Fig. 5.9) but its fat content decreases after 48 hours, together with that of the entire seedling, while the changes in the fat content of the endosperm and axis are quite small (Fig. 5.9). The overall trend is therefore similar, movement from the endosperm to the embryonic axis. Soluble carbohydrates increase in all parts of the seedling after about 48 hours of germination.

The overall changes in the composition of rice seeds during germination is illustrated in Table 5.1. Again a loss of starch is observed together with increases in free sugars, soluble amino N and soluble protein and a decrease in RNA. The loss in dry weight begins after only about 4 days, indicating an initial slow rate of metabolism. Other investigations indicate that the loss in starch occurred in the endosperm, but no

Table 5.1—Changes in Composition of Rice during Germination in the Dark
(From data of Palmiano and Juliano, 1972)

| Period of germination days | Dry weight mg | Starch mg | Free sugars mg | Crude protein mg | Soluble amino N $\mu$g | RNA $\mu$g | Soluble protein $\mu$g |
|---|---|---|---|---|---|---|---|
| 0 | 18·4 | 16·2 | 0·15 | 1·36 | 2·18 | 26 | 258 |
| 2 | 19·6 | — | — | 0·83 | — | 14 | 276 |
| 3 | 17·0 | 13·9 | 0·37 | 0·82 | 9·79 | 11 | 268 |
| 4 | 17·0 | 12·4 | 0·77 | 0·70 | 14·25 | 11 | 296 |
| 5 | 12·6 | 10·8 | 1·14 | 0·64 | 15·80 | 11 | 304 |

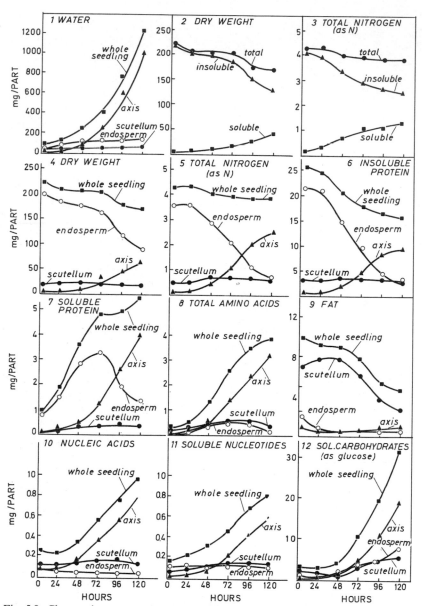

Fig. 5.9. Changes in content of various components in different parts of *Zea mays* during germination.
(From Ingle *et al.*, 1964)

accumulation occurred in the root or shoot. Dry weight increase was more rapid in the shoot than in the root, since in rice the coleoptile grows out before the shoots (Fukui and Nikuni, 1956).

Yocum (1929) grew wheat in soil and analysed the composition of both the entire seed or seedling and, after separation into different organs was possible, also the composition of the endosperm, root and plumule. Table 5.2 shows some of the results

Table 5.2—Changes in Composition of Wheat Seeds during Germination and Growth
The table illustrates the change in storage materials during germination and growth. The data show both the composition of the entire seed or seedling and also the composition of the endosperm, root and plumule after different times of germination. The results are given as weight of substances in mg per 100 seeds or parts of seedlings
(Compiled from Yocum, 1925)

| Time in days | Plant part | Dry weight | Ether extract (lipids) | Reducing sugars | Total sugar | Dextrin | Starch | N |
|---|---|---|---|---|---|---|---|---|
| 0 | Original Seed | 2685 | 66·9 | 0 | 53·9 | 43·5 | 1781·0 | 59·1 |
| 1 | Seed | 2708 | 63·1 | 0 | 44·7 | 82·0 | 1621·3 | 54·4 |
| 2 | Seed | 2593 | 57·6 | 22·8 | 137·2 | 111·0 | 1079·0 | 46·2 |
| 3 | Seed | 2544 | 51·0 | 83·1 | 121·4 | 81·3 | 1343·5 | 48·8 |
| 6 | Seedling | 2476 | 45·9 | 234·8 | 465·8 | 120·5 | 472·1 | 49·6 |
| 9 | Seedling | 2383 | 95·3 | 117·3 | 208·9 | 86·5 | 117·5 | 63·3 |
| 12 | Seedling | 2031 | 94·7 | 15·6 | 37·1 | 19·9 | 20·2 | 54·3 |
| 3 | Plumule | 110 | 2·1. | 3·1 | 9·2 | 3·3 | 2·6 | 5·9 |
| 6 | Plumule | 383 | 8·7 | 13·6 | 20·4 | 9·6 | 2·9 | 14·8 |
| 9 | Plumule | 875 | 54·8 | 13·6 | 26·5 | 11·7 | 7·0 | 37·7 |
| 12 | Plumule | 1054 | 68·3 | 0 | 4·1 | 3·0 | 0 | 34·8 |
| 3 | Root | 103 | 2·6 | 3·6 | 6·3 | 2·5 | 1·5 | 5·4 |
| 6 | Root | 286 | 6·1 | 8·5 | 14·8 | 5·2 | 0·6 | 7·5 |
| 9 | Root | 469 | 10·7 | 11·7 | 15·5 | 6·8 | 1·5 | 12·9 |
| 12 | Root | 691 | 10·2 | 7·2 | 12·5 | 7·8 | 2·3 | 15·9 |
| 3 | Remaining Seed (endosperm) | 2332 | 46·3 | 76·5 | 105·9 | 75·4 | 1339·5 | 37·6 |
| 6 | (endosperm) | 1807 | 31·2 | 212·7 | 430·7 | 105·7 | 468·0 | 27·3 |
| 9 | (endosperm) | 839 | 29·7 | 102·1 | 167·0 | 60·1 | 109·0 | 12·8 |
| 12 | (endosperm) | 287 | 16·3 | 8·4 | 20·5 | 8·9 | 17·9 | 3·6 |

obtained. It will be noted that starch decreased continuously in the endosperm as well as in the whole plant. Dextrins also disappeared from the endosperm, although on the first day some dextrins are formed from other substances. Fats also decreased during the first few days, but later fats are reformed. The resynthesis of fats and the appearance of reducing and other sugars is presumably the result of transformations of the disappearing starch. Nitrogen also leaves the endosperm and appears in other parts of the seedling.

Oil-containing seeds have been investigated by Yamada (1955). He showed the disappearance of lipids in the various parts of the seeds of the castor bean during germination. Total lipids decreased both in the endosperm and the cotyledons, as did the total amount of fatty acids as germination proceeds (Fig. 5.10). In contrast carbohydrates began to appear in the endosperm as the fats disappear, showing accumulation up to 4 days. After this, carbohydrates disappeared again from the

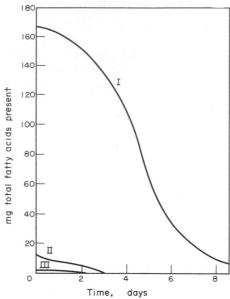

Fig. 5.10. Disappearance of fatty acids in various parts of the seed during germination of
castor bean seed.
I—Endosperm; II—Cotyledons; III—Embryo
(Yamada, 1955)

endosperm and appeared in the hypocotyl. This is true both for reducing and
non-reducing sugars (Fig. 5.11). It appears that fats are converted  to sugar in the
endosperm itself, and later the carbohydrates are transferred to the embryo.

A detailed study of the utilization of lipids in Douglas fir seeds during germination
showed a decrease in total lipids. Glycerides were utilized while certain phospholipids
increased slowly at first and then more rapidly. The relative distribution of short and
long chain fatty acids changed considerably in the free fatty acid fraction (Ching, 1963,
1966). It is obvious that the various lipid fractions undergo different metabolic fates
during germination. This is probably true not only for Douglas fir, but for all lipid
containing seeds.

In *Trigonella foenum-graecum* galactomannans are a major reserve substance.
Their breakdown begins 18 hours after radicle emergences. During breakdown the
galactose-mannose ratio changes. As the galacto-mannan is broken down the starch
content of the cotyledons begins to rise sharply (Reid, 1971).

Semenko (1957) investigated the changes in nucleic acids occurring during
germination of wheat and oat seeds. The amount of RNA and DNA in the wheat but
not the oat endosperm decreased and it increased in the embryo and subsequently in
the seedling. The total amount of nucleic acids accumulating in the seedling after 10
days is greater than the initial amount present in the endosperm, despite the fact that
the endosperm still contained some nucleic acids (Table 5.3).

Overall increases in the RNA and DNA content have been reported during the
germination of many seeds, including lettuce, rye and castor beans. In the case of

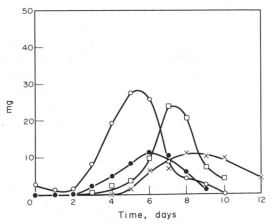

Fig. 5.11. Changes in reducing and non-reducing sugars during the germination of castor beans

O——O   Non-reducing sugar in endosperm
●——●   Reducing sugars in endosperm
□——□   Non-reducing sugar in hypocotyl
×——×   Reducing sugars in hypocotyl
(Yamada, 1955)

Table 5.3—Changes in Nucleic Acid Content in Endosperm and Embryo of Seedling of Germinating Wheat and Oats
The results are given as mg pyrophosphate per 100 endosperms or seedlings
(After Semenko, 1957)

| | | Nucleic acid content | | |
| | | Endosperm | | Seedling |
| Time in days | RNA | DNA | RNA | DNA |
| --- | --- | --- | --- | --- |
| | 0 | 3·28 | 1·85 | 2·15 | 1·02 |
| Wheat | 2 | 2·89 | 1·64 | 2·76 | 1·22 |
| | 6 | 1·28 | 1·57 | 5·76 | 2·03 |
| | 10 | 1·19 | 0·98 | 7·17 | 4·85 |
| | 0 | 2·14 | 0·45 | 0·97 | 0·59 |
| Oats | 6 | 0·44 | 0·62 | 1·15 | 1·38 |
| | 10 | 0·18 | 0·91 | 3·53 | 1·86 |

complex molecules such as the nucleic acids care must be taken in interpreting the decrease in content in one part of the seed and its increase in another part as transport of the compound from one part to another. Probably breakdown occurs in the storage tissue, cotyledons or endosperm and synthesis in the growing tissues, in which active cell division and elongation occur. A reorganization of the different forms of nucleic acids also occurs during germination.

An interesting approach to the study of mobility of seed components is that of McConnell (1957). McConnell obtained radioactive wheat seeds by feeding the parent plant with acetate, labelled in carbon 1 or carbon 2 with $C^{14}$. He then fractionated the seeds and determined the specific activity of various seed fractions. After germinating the seeds for 5–7 days, he again determined the specific activities of the various fractions in the kernel, roots and stem of the seedling. He was able to show that a considerable part of the carbon, 17 per cent, was lost as carbon dioxide during respiration. In the remaining $C^{14}$ he found a redistribution of labelled carbon, showing increases in specific activity in some fractions and decreases in others, again indicating the mobility of the constituents of the seed and their movement from the seed kernel to the stem and roots, which he analysed separately. He was also able to show that part of the protein was being respired, apparently via glutamic acid.

In most seeds only the overall changes in certain components have been studied. Most of these clearly indicate which materials in the seeds show a loss or gain.

## II.  Breakdown and Metabolism of Storage Products and Synthesis of Nucleic Acids and Proteins

From the discussion on the changes in storage products it can be concluded that some of them undergo very marked metabolic changes. Not necessarily those substances which are present in the greatest amounts are broken down first. The metabolic changes occurring in the early stages of germination are the result of the activity of various enzymes. The hydrolytic enzymes and transfer enzymes are either present in the dry seed or very rapidly become active as the seed imbibes water. In view of the importance of some of these enzymes, the reactions which are catalysed by them will be briefly discussed.

Generally, enzymes breaking down starch, proteins, hemi-celluloses, polyphosphates, lipids and other storage materials, rise in activity fairly rapidly as germination proceeds, although there is no reason to think that these changes are the direct cause of the actual germination process. Changes in the activity of various enzymes have been studied in many seeds but the same general trend is observed.

### 1.  *Carbohydrates*

Starch is normally broken down by amylases. However, the exact path by which these amylases function is different in different seeds. Usually seeds contain both amylose and amylopectin. Dry seeds contain primarily $\beta$-amylase which detaches units of maltose from the starch molecule. In the case of attack on amylopectin dextrins are formed, while when amylose is attacked the molecule is broken down more or less completely. In most seeds, as germination proceeds, other enzymes attacking starch are liberated. More particularly $\alpha$-amylase is formed, but in addition debranching enzymes also appear. Small amounts of $\alpha$-amylase are present in dry kernels of wheat. Phosphorylases also appear to be present, both in dry seeds and in germinated ones. Obviously the precise course of starch breakdown in any given seed will be determined by the relative amounts of those enzymes present, as well as by the ratio of the two main forms of starch, amylopectin being broken down preferentially.

The result of the hydrolysis of starch by $\beta$-amylase, maltose, rarely accumulates as such in the seeds. Maltose is further broken down to glucose by maltase. Attack on starch by $\alpha$-amylase results in a mixture of sugars, including maltose and glucose. The maltose again is usually broken down further. The enzymes involved in breakdown and synthesis of starch in cereals have been reviewed by Marshall (1972).

*Zea mays* seeds contain only $\beta$-amylase in the dry seed. The rise in amylase activity in the seed during germination is primarily in $\alpha$-amylase which, when amylolytic activity is at its peak, accounts for 90 per cent of total amylolytic activity of the endosperm. The $\alpha$-amylase originates in the scutellum and is secreted into the endosperm, while $\beta$-amylase appears to form only in the endosperm (Dure, 1960). In barley seeds it was found that the rise in $\alpha$-amylase activity of the endosperm depended on the presence of the embryo. Removal of the latter resulted in a drop in amylase activity (Kirsop and Pollock, 1957).

Formation of $\alpha$-amylase in the barley endosperm is controlled by gibberellic acid, which is at least initially secreted by the scutellum. However, the scutellar secretion is in turn controlled by the axis (Radley, 1968). Enzyme formation occurs by *de novo* synthesis in the aleurone layer (Varner *et al.*, 1965). The significance of this control will be discussed in the following chapter. In other cases formation of hydrolytic enzymes may be controlled by other growth substances, such as cytokinins. In some cases, at least, the formation of hydrolytic enzymes appears to be a release reaction, for example $\beta$-1·3-glucanase in barley aleurone. A starch debranching enzyme is released from a particulate fraction in peas by an activation reaction. Breakdown of starch in peas was shown by Swain and Dekker (1966) to follow the pathway:

$$\text{starch} \xrightarrow{\quad} \text{soluble oligosaccharides} \xrightarrow{\quad} \text{maltose} \xrightarrow{\quad} \text{glucose}$$
$$\quad\alpha\text{-amylase} \qquad\qquad\qquad\qquad\qquad \beta\text{-amylase} \quad \alpha\text{-glucosidase}$$

However, in pea seeds, phosphorylase rises very rapidly in the cotyledons during 4 days of germination, so that phosphorolytic breakdown probably also occurs. The changes in the enzymes involved in starch breakdown during the germination of peas are illustrated in Fig. 5.12. Low activity of $\beta$-amylase was present in the dry seed, but no $\alpha$-amylase activity could be detected. In how far germination and rise in enzyme activity are causally related is at best doubtful as germination preceded the rise in the activity of the various enzymes.

The changes in the carbohydrates of barley during germination have been studied in detail because of their importance in the malting process. Glucose and fructose rise very considerably up to 6 days' germination at 21°C and then begin to fall again. After 6 days the seedlings were 5–7 cm long. Other sugars also showed marked changes. Thus maltose rose from 1 mg per gram dry seeds to more than 55 mg after 7 days of germination. Almost as great an increase occurred in oligosaccharides containing more than three hexose residues. Sucrose showed far smaller and less regular increases, as did glucodifructose. Raffinose and maltotriose stayed more or less steady for the first 5 days of germination and then rose steeply, increasing five-fold in the next 2 days. Apparently sucrose, raffinose, glucodifructose and fructosans are associated with respiration (McLeod, Travis and Wreay, 1953). Raffinose metabolism in barley was studied in greater detail by McLeod (1957). Raffinose was absent from

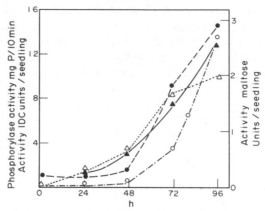

Fig. 5.12. Development of enzymes involved in starch breakdown in peas during germination.
● β-amylase (per seedling)
△ amylopectin-1,6-glucosidase (per seedling)
○ α-amylase (per seedling)
▲ phosphorylase (per 30 cotyledons).
(Redrawn from data of Shain and Mayer, 1968 and Swain and Dekker, 1966)

the endosperm of the barley, but accounted for 9 per cent of the dry weight of the embryo. Raffinose was rapidly utilized by the embryo under normal conditions, but at the same time the sucrose content of the seedling increased. In isolated embryos raffinose metabolism was retarded. No changes in either sucrose or raffinose occurred during the first 24 hours of germination if the seeds were kept immersed in water, i.e. steeped as in the malting process. Their metabolism seems to be closely connected to aerobic processes.

Since the endosperm of the cereal seeds is connected to the seedling by the scutellum, the latter is of some interest in considering the metabolism of such seeds during germination. Edelman *et al.* (1959) studied the function of the scutellum in considerable detail. They were able to show that glucose is removed from the endosperm, converted to sucrose in the scutellum and transported as such to the embryo. The scutellum always has a low hexose and high sucrose content, the reverse being true for the endosperm and for the seedling. Even the isolated scutellum can readily form sucrose from hexose. Sucrose was formed by a complex mechanism. Glucose is phosphorylated in the six position in the presence of ATP. Part of the glucose-6-phosphate formed is converted to fructose-6-phosphate (F-6-P) and part to glucose-1-phosphate. The glucose-1-phosphate is converted to uridine diphospho-glucose (UDPG), in the presence of uridine triphosphate (UTP). Sucrose is then formed by the condensation of UDPG and F-6-P. The enzymes necessary for all these reactions could also be demonstrated in the scutellum.

Other enzyme systems concerned with carbohydrate metabolism were shown by Lechevallier (1960) in *Phaseolus vulgaris* seeds. In these α-galactosidase is present in appreciable amounts in the embryo and of low activity in the cotyledons of the dry seeds. During germination its activity falls in the embryonic axis and rises in the cotyledons. An interesting feature of metabolism in *Phaseolus* seeds is the formation of malonic acid during germination. In the dry seeds this is absent or present in trace amounts, while after 5 days germination appreciable quantities are found in the embryonic axis (Duperon, 1960). In the same seeds other organic acids were also metabolized. Citric acid decreases during germination while malic acid accumulates, marked quantitative changes again occurring after 5 days. However, in *Zea mays*, citric, malic and aconitic acid all increase, though at different rates, during germination. Other tricarboxylic acid cycle intermediaries were found, if at all, only in very small amounts in a number of seeds (Duperon, 1958).

Sucrose is often present in dry seeds in small amounts or is formed as a result of raffinose breakdown. The presence of invertase has been demonstrated in a number of germinating seeds, for example barley (Prentice, 1972) and lettuce (Eldan and Mayer, 1974). It arises during germination and could at least partly account for sucrose breakdown. In addition, however, it is probable that part of the sucrose is metabolized by glycosyl transfer reactions (Pridham *et al.*, 1969).

Generally speaking it may be said that most of the enzymes involved in the breakdown and interconversion of the carbohydrates became active during germination, most by *de novo* synthesis, some by activation or release.

## 2. *Lipids*

Fats and oils are broken down in the first instance by the action of lipases. Lipases are rather non-specific esterases which cleave the bond between the fatty acids and the glycerol which esterifies them. Normally neither of the breakdown products of hydrolysis of lipids accumulates in the seeds. The fate of the glycerol which is formed is not known with certainty. It seems, however, to become part of the general carbohydrate pool present in the seed and as such becomes available for various processes including respiration. Thus enzyme systems have been shown in *Arachis* cotyledons which convert glycerol to glycerol phosphate which is then converted to triose phosphate. This can then be either converted to pyruvic acid or to sugars (Stumpf and Bradbeer, 1959). The fatty acids formed following lipase action accumulate in small amounts. The bulk of the fatty acids is, however, broken down further by one of a number of reactions. The fatty acids may be broken down by the process of $\beta$-oxidation, resulting in the cleavage of two carbon units in the form of acetyl, which can enter the tricarboxylic acid cycle. This reaction requires both CoA and ATP. $\beta$-oxidation has been demonstrated in extract of various seeds (Rebeiz *et al.*, 1965). At least in castor bean endosperm the $\beta$-oxidation activity is entirely associated with the glyoxysomes (for discussion of glyoxysomes see later). The fatty acids may also be broken down by $\alpha$-oxidation. In this process the fatty acid is peroxidatively decarboxylated and carbon dioxide formed. The long chain aldehyde is oxidized to the corresponding acid by a reaction linked to NAD. In many seeds disappearance of fats is accompanied by the appearance of carbohydrates. This reaction apparently proceeds as follows. The fatty acids undergo $\beta$-oxidation. The acetyl CoA formed is converted to malate via the glyoxylate cycle. The malate thus formed is converted to carbohydrate by a number of reactions. All these reactions have been shown to occur in the cotyledons or endosperm of fat-containing seeds such as soybean, castor beans and groundnuts (*Arachis*). Another enzyme which is believed to play a part in fatty acid oxidation is lipoxidase. This enzyme, which also occurs in seeds, is supposed to break the fatty acid chain into two smaller parts by peroxidative attack at a double bond. The precise function of this system is at present in doubt.

The fat content of seeds changes during germination as already mentioned. The detailed course of fat metabolism has been followed by Hardman and Crombie (1958) and Boatman and Crombie (1958) in two different seeds, *Citrullus vulgaris* and *Elaeis guineensis*. In *Citrullus* seeds there is a rapid breakdown of lipids both in the cotyledons and in the rest of the seed. These lipids are largely used in respiration and do not seem to be converted to carbohydrates. Certain differences in the course of lipid metabolism of seeds grown in the dark and in the light were found. Thus in the light all the fats were broken down at an equal rate. In the dark, however, there seems to be a greater disappearance of linolenic acid than in the light. Another difference in fat metabolism in the light and the dark was the rate of appearance of phosphatides. In the light they later formed continuously for 15 days, while in the dark they formed for the first 7 or 8 days and then their amount decreased again. However, photosynthesis begins in these seedlings fairly quickly. The difference observed may therefore be due to processes of assimilation rather than an essential difference in behaviour of germination in light and dark.

A rather different situation was found for *Elaeis* seeds. In these seeds the lipids are located in the endosperm, which is invaded by an haustorium during germination. Free fatty acids accumulate in the endosperm but not in the haustorium. Although lipids are found in the haustorium, they appear in the esterified form. It seems that free fatty acids are transferred from the endosperm to the haustorium and are immediately re-esterified there. In *Elaeis* seeds the bulk of the lipids is lost during germination, apparently in respiration, and little or no conversion to carbohydrate occurs. All fatty acids are metabolized at about the same rate, but saturated fatty acids disappear a little more rapidly than unsaturated ones.

Lipase has been frequently claimed to be absent from oil-containing seeds. These results can probably be ascribed to the insolubility of lipases. In order to show the presence of lipase activity it is usually necessary to defat the seeds and then activate the enzyme. Such activation is done by treating the defatted powder with dilute acetic acid or with a solution of calcium chloride. This activation may liberate the enzyme from an inactive form. The usual way of assaying lipase activity is to use the defatted powder of the seeds as an enzyme source and an emulsion of a suitable oil as a substrate. As already stated, lipases are rather non-specific in their action. As an example, a crude lipase prepared from *Brassica campestris* (rape seed) was able to hydrolyse tributyrin, triacetin and triolein as well as natural olive, sunflower and rape oils. However, triolein was attacked more slowly than any of the other substrates. Triacetin was attacked at about half the rate of tributyrin (Wetter, 1957). The author points out that some sort of specificity in the hydrolytic attack on the different oils seems to be involved and that the observed differences were not merely caused by differences in the solubility or degree of dispersion of the oils.

The lipase activity of seeds changes with germination time. It is frequently found

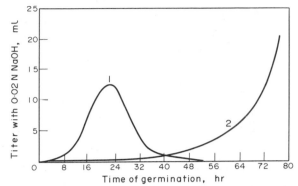

Fig. 5.13. Lipase activity in castor bean seeds during germination.
1—Neutral lipase in the embryo; 2—Neutral lipase in the endosperm.
(Yamada, 1957)

that more than one lipase is active. These are differentiated in various ways including the pH at which they function. In this way it is possible to distinguish between neutral and acid lipases. Figure 5.13 shows the changes in activity in the neutral lipase in the endosperm and the embryo of germinating castor beans (*Ricinus communis*). Yamada (1957) investigated the fat metabolism of these seeds. He found, in addition to the neutral lipase, an acid lipase which is already present in the endosperm of the dry seed.

Lipases and lipid metabolism have been investigated in considerable detail in seeds of Douglas fir (Ching, 1968) in which the enzyme appears to be associated with the fat bodies themselves. In other cases the lipases have been shown to be associated with spherosomes. Even in seeds with a low lipid content, such as wheat, lipid metabolism develops quite rapidly during germination in all parts of the seed and seedling (Fig. 5.14). Nevertheless, breakdown of lipids was not always correlated with lipase activity.

Special consideration in lipid metabolism must be given to the glyoxylate cycle. The relation of this metabolic pathway to the tricarboxylic acid cycle on the one hand and the conversion of fats to carbohydrates is illustrated in Fig. 5.15a. The important feature of this cycle is that acetyl CoA, which is formed from pyruvate condenses with glyoxylic acid to form malate in the presence of the enzyme malate synthetase and ATP. The glyoxylate arises from isocitric acid, with the concomitant formation of succinic acid by the action of isocitrate lyase. The succinic acid can enter the tricarboxylic acid cycle, as can the malate which is formed from the glyoxylate. Evidence for the operation of this pathway has been brought for both *Arachis* and *Ricinus communis* seeds (Kornberg and Beevers, 1957; Marcus and Velasco, 1960; Yamamoto and Beevers, 1960). It appears that in both castor beans and peanuts the chief function of the glyoxylic acid cycle is the conversion of fats to carbohydrates. This conversion is effected via the oxidation of the fatty acids by $\beta$-oxidation (as previously mentioned) and formation of acetyl CoA, which is in turn converted to malate. The malate is converted to carbohydrates via oxaloacetate and phosphopyruvate, a process which requires ATP. At the same time it seems that $\alpha$-ketoglutarate in the tricarboxylic acid cycle can be completely by-passed due to the operation of the

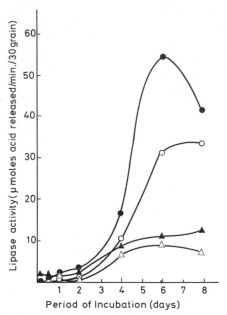

Fig. 5.14. Lipase activities in the tissues of germinating wheat grain.

● starchy endosperm
○ bran
▲ embryo axis
△ scutellum

(From Tavener and Laidman, 1972)

glyoxylic acid cycle (Marcus and Velasco, 1960). An alternative point of entry of two carbon units into the respiratory mechanism is provided by this means. The activity of the enzymes involved in this cycle has been shown to increase during germination of *Arachis* seeds, as shown in Table 5.4. It will be seen that both malate synthetase and isocitrate lyase increase during germination and are completely absent in the dry seed.

In most cases all the enzymes of the glyoxylate cycle are present within special

Table 5.4—Changes in Activity of the Glyoxylate Cycle during
Germination of Peanuts
Enzyme activity given as $\mu$ moles glyoxylate formed (isocitritase) or
removed (malate synthetase)
(From Marcus and Velasco, 1960)

| Time in days | Length of radicle (mm) | Enzyme activity | |
| | | Isocitritase | Malate synthetase |
| --- | --- | --- | --- |
| 0 | | — | |
| 1 | 0 | 0 | 0 |
| 2 | 2 | 7·9 | 10·2 |
| 3 | 16 | 20·4 | 24·6 |
| 4 | 35 | 30·0 | 50·2 |
| 5 | 55 | 42·3 | 69·4 |

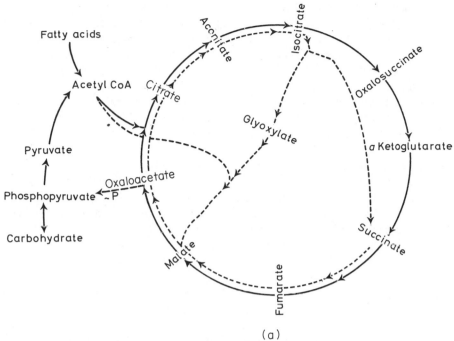

(a)

Fig. 5.15a. The relation between the tricarboxylic acid cycle (solid line) and the glyoxylate cycle (broken line) and the possible way of conversion of fat to carbohydrate.

subcellular organelles—the glyoxysomes (Fig. 5.15b). These were first described by Breidenbach and Beevers (1967). The glyoxysomes are particles of slightly greater density than the mitochondria, from which they can be separated by differential centrifugation. Isocitrate lyase and malate synthetase appear to be associated solely with these particles, although soluble activity of the lyase has been reported in some tissues. Glyoxysomes are formed during germination and disappear again as germination proceeds and the lipid reserve is broken down. Enzyme activity is under quite complex control, but most of the enzyme appears during germination by *de novo* synthesis (Mayer and Shain, 1974). The occurrence of glyoxysomes has now been reported in a sufficient number of lipid containing seeds so that their presence may be regarded as general in all those tissues in which gluconeogenesis from lipids takes place (Richardson, 1974).

### 3. *Proteins*

As already stated, all seeds contain a certain amount of protein. In Figs. 5.5 and 5.7 were shown the breakdown of protein in cotyledons of beans on the one hand, and the appearance of new proteins in other parts of the seedling on the other hand. Other nitrogenous compounds also appear as germination proceeds.

The presence of proteases and peptidases has been shown in many seeds, although most attention has been given to these enzymes in barley, because of their importance in the malting process. Some of the proteins and peptidases are present in the dry seed, while others appear during germination. In most cases the enzymes having proteolytic activity are soluble and are present or develop in the storage organs i.e. in the cotyledons or the endosperm. The types of proteolytic enzymes and peptidases occurring in germinating seeds do not seem to differ fundamentally from those occurring in other plant tissues, although in a few cases enzymes with special characteristics have been reported. The appearance of proteolytic enzyme activity appears to be under hormonal control of the axis or embryo. Thus in barley the appearance of enzyme activity is regulated by gibberellic acid, while in *Cucurbita maxima* cytokinin regulates enzyme activity. Such hormonal control is not, however, always observed and seems to be absent in pea cotyledons.

The proteolytic enzymes of germinating seeds show great diversity both as regards their specificity for peptide linkages, their pH optima and their reaction to inhibitors. The same also applies to the peptidases. For example in barley eight distinct peptidases are present, as well as three different proteases. In pea seeds at least two peptidases and two proteases are present, and in lettuce four different proteolytic enzymes could be detected. In many seeds endogenous inhibitors of trypsin and chymotrypsin are present. These inhibitors have been studied in some considerable detail (Vogel *et al.*, 1968) because of their importance when the seeds are ingested by animals. Their exact function in the germination process is still debatable. They may act by regulating proteolytic enzyme activity during germination or they may be simply a relic from the period of seed development, when they may have prevented the decomposition of newly formed storage proteins. The protease inhibitors are not equally distributed among different parts of the seed or seedling.

The change in proteolytic enzyme activity of germinating lettuce is illustrated in Fig. 5.16. As can be seen two enzymes of different pH optima increase during germination while a third enzyme disappears, due to its interaction with an endogenous trypsin inhibitor. In soya bean proteinase activity was studied by Tazakawa and Hirokawa (1956). Activity rose much more rapidly and earlier in the cotyledons than in the axis (Fig. 5.17). The extent to which the proteolytic enzymes degrade seed protein has not been investigated in most cases. It may, however, be assumed that the proteolytic enzymes do in fact degrade the storage protein to soluble nitrogenous compounds, which in turn are utilized by the various parts of the seedling.

Although many questions are still open with regard to detailed mechanism of protein metabolism during germination, there is some information about the fate of the breakdown products. Usually there is little change in the total nitrogen content of the seed or seedling during germination, although slight losses may occur, especially due to leaching out of nitrogenous substances. Nitrogen appears to be very carefully conserved. In place of the protein broken down there appear free amino acids and amides. The mechanism by which these substances are formed, especially if seeds are germinated in the dark in water, has been the subject of extensive research. In most cases only the breakdown of proteins and the simultaneous appearance of amides and amino acids has been followed. Table 5.5 shows some of these changes in *Phaseolus mungo* seeds.

The early work on nitrogen metabolism both by Paech (1935) and Prianishnikov

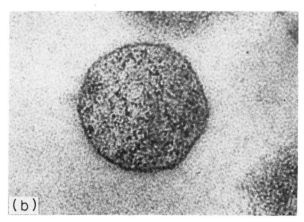

Fig. 5.15b.  Electron micrograph of glyoxysome from castor beans. ×60000
(Courtesy of Dr. Y. Shain)

Table 5.5—Changes in Protein Content and Other Nitrogenous Compounds in *Phaseolus Mungo* Seeds
(From Damodaran *et al.*, 1946)
The results are given as percentage of total nitrogen

| Age of seedlings in days | 1 | 4 | 7 | 10 | 13 | 16 |
|---|---|---|---|---|---|---|
| Protein N (Extractable) | 83·0 | 35·0 | 20·0 | 9·0 | 13·0 | 7·0 |
| Ammonia N | 0·3 | 0·9 | 1·2 | 0·8 | 1·1 | 2·5 |
| Asparagine amide N | 0·7 | 9·3 | 11·3 | 12·7 | 12·5 | 11·7 |
| Glutamine amide N | 0·3 | 1·1 | 1·4 | 2·3 | 2·9 | 2·5 |
| Amine N | 3·2 | 15·4 | 23·7 | 23·7 | 22·5 | 20·8 |
| mg Total N/300 seedlings | 534·5 | 486·2 | 511·2 | 517·2 | 499·6 | 505·5 |

germination in the dark, proteins are broken down to amino acids. Part of these amino acids are oxidatively de-aminated and the carbon skeleton enters various respiratory and carbon cycles. The ammonia formed by deamination is detoxicated by the process of amide formation. The chief amides formed are glutamine and asparagine, depending on the plants.

Not all amino acids are deaminated in this way. Part of them are utilized for the synthesis of proteins in the actually growing parts of the seedling. Soluble nitrogen already present in the seedling may also be utilized during germination. Thus Egami *et al.* (1957) showed that the small amounts of nitrate present in *Vigna* seeds disappear as germination proceeded; they also proved the existence of nitrate reductase system in the seedlings. Yamamoto (1955) showed in the same seeds that asparagine present in the cotyledons disappears and instead appears in the hypocotyl and plumule. On the basis of these various findings the effect of feeding ammonia and sugars in protein metabolism can also be understood. In the case of feeding ammonia, protein breakdown is reduced, because external nitrogen can be used to form new protein. In the case of sugars, the sparing action is chiefly due to the provision of respiratory substrates, which would otherwise be provided by breakdown of protein and de-amination of the amino acids formed.

The mechanism of amide formation is fairly well established as far as glutamine is concerned. Glutamine is formed from glutamic acid and ammonia, in the presence of the enzyme glutamine synthetase and ATP, the reaction being energy-requiring. The precise biochemical mechanism has also been established. As far as asparagine formation is concerned the situation is less clear. However, Yamamoto (1955) showed that in *Vigna* hypocotyl, asparagine is formed if the hypocotyls are fed with aspartic acid, ammonia and ATP. The enzyme responsible for the reaction has, however, never been isolated. The formation of asparagine may involve $\beta$-cyanoalanine as an intermediate.

In addition to enzyme systems causing the synthesis of amides, germinating seeds usually contain enzymes causing hydrolysis of the amide bond. These are glutaminase and asparaginase, whose precise function is unknown. Far more interesting are those enzymes which can transfer amino groups from the amide to some keto acid which results in amino acid formation.

Another type of transfer enzyme concerned with amino acid metabolism is constituted by the transaminases. These enzymes transfer amino groups from amino acid to keto acid (see for example, Fowden, 1965). The presence of transaminases in a variety of seeds has been shown by Smith and Williams (1951). They found a marked increase in the activity of the transaminases transferring amino groups from alanine or aspartic acid to $\alpha$-ketoglutaric acid (Table 5.6). It is probable that the same enzymes are also carrying out the reverse reactions. In most cases glutamic-aspartic transaminase increased more rapidly than the glutamic-alanine transaminase. However, in pea the reverse was the case, and in corn both enzymes seemed to increase at about the same rate. These authors also showed that no fixed relation existed between increase in protein in the seeds and increase in transaminase activity, and concluded that there was no evidence to show that the two processes were directly connected.

Evidence for the presence of enzymes catalysing the reverse reaction from glutamine to pyruvic or oxaloacetic acid in wheat germination was brought by Cruickshank and Ishwerwood (1958).

Albaum and Cohen (1943) followed the transamination reaction glutamic-oxaloacetic acid and its reverse reaction in oat embryos. The former action took place at three times the rate of the latter. Glutamic-oxalo acetic acid transaminase activity of the embryo expressed per unit of protein increased steadily during germination. Expressed per unit of dry weight it decreased, probably because of the large increase in dry weight of the embryos. In contrast to the results of Smith and Williams they found good correlation between increase in protein, transaminase activity and soluble nitrogen content. The changes are illustrated in Fig. 5.18.

Little information is available with regard to the importance of these reactions in amino acid and protein synthesis in the germinating seed. It must be remembered that amide formation is almost entirely absent in seedlings germinated in the light. In such

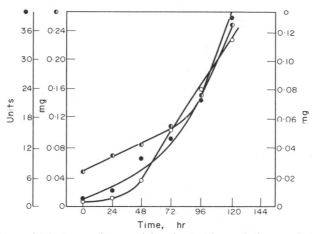

Fig. 5.18. Changes in the transaminase activity of oat embryos during germination together with changes in solubles and protein nitrogen.
◐ Protein as mg N
○ Soluble nitrogen as mg N
● Transaminase activity as units of activity
(After Albaum and Cohen, 1943)

Table 5.6—Changes in Glutamic-aspartic and Glutamic-alanine Transaminases and in Protein Nitrogen during Germination of Various Plant Seeds. Enzyme activity is expressed in units/embryo and protein nitrogen as mg protein/embryo. In all cases the embryo after removal of endosperm or cotyledons was used
(Smith and Williams, 1951)

| Time in hours | 24 | | | 48 | | | 72 | | | 96 | | |
|---|---|---|---|---|---|---|---|---|---|---|---|---|
| Seeds | Glu-Al | Glu-Asp | Protein N | Glu-Al | Glu-Asp | Protein N | Glu-Al | Glu-Asp | Protein N | Glu-Al | Glu-Asp | Protein N |
| Waxbean | 0·03 | 0·10 | 0·11 | 0·11 | 0·16 | 0·16 | 0·15 | 0·28 | 0·26 | 0·17 | 0·48 | 0·32 |
| Pea | 0·03 | 0·06 | 0·09 | 0·17 | 0·13 | 0·12 | 0·28 | 0·18 | 0·15 | 0·37 | 0·23 | 0·16 |
| Barley | 0·08 | 0·09 | 0·03 | 0·14 | 0·26 | 0·04 | 0·24 | 0·56 | 0·08 | 0·42 | 0·88 | 0·11 |
| Corn | 0·09 | 0·10 | 0·08 | 0·17 | 0·17 | 0·11 | 0·20 | 0·28 | 0·16 | 0·44 | 0·53 | 0·32 |
| Oats | 0·05 | 0·14 | 0·02 | 0·10 | 0·18 | 0·02 | 0·17 | 0·44 | 0·04 | 0·20 | 0·67 | 0·07 |
| Squash | 0·06 | 0·04 | 0·08 | 0·07 | 0·06 | 0·08 | 0·08 | 0·13 | 0·08 | 0·13 | 0·43 | 0·09 |
| Pumpkin | 0·04 | 0·06 | 0·04 | 0·07 | 0·09 | 0·06 | 0·16 | 0·27 | 0·08 | 0·28 | 0·43 | 0·10 |

seedlings, especially when the seeds are germinated in the soil or in nutrient solution, protein synthesis is from external nitrogen taken up by the seedling and from the carbon skeleton formed during photosynthesis. Undoubtedly the key processes in such seedlings are the *de novo* synthesis of amino acids and proteins and not protein breakdown and amide formation. In amino acid synthesis transamination reactions probably play an important role.

A little evidence about amino acid synthesis is available. Virtanen *et al.* (1953) were able to show the synthesis of homoserine in pea seeds during the first 24 hours of germination. Homoserine was not present in the dry seed, either in the free state or in protein of the seed. Thus during germination there occurred the rapid synthesis of an entirely new amino acid. The function of this amino acid was, however, not studied. A general rise in the amino acid content of lettuce seeds during germination was shown by Klein (1955). This is shown in Table 5.7. Such changes appear to be quite general during germination.

Table 5.7—Changes in the Amino Acid Content of Lettuce Seeds
during Germination
(Klein, 1955)
The amino acid content is given as $\gamma$-amino N per gram initial
dry seeds

| Amino acid | Time of germination in days | | | |
|---|---|---|---|---|
| | 0 | 1 | 2 | 3 |
| Alanine | 5 | 30 | 80 | 220 |
| Threonine | 5 | 20 | 40 | 190 |
| Leucine | 20 | 20 | 60 | 280 |
| Serine | 30 | 30 | 60 | 250 |
| $\gamma$-amino butyric acid | 5 | 5 | 15 | 25 |
| Lysine | 15 | 5 | 20 | 40 |
| Tryptophane | 5 | 5 | 2 | — |
| Glutathione | 10 | 0 | 0 | 20 |
| Aspartic acid | 40 | 35 | 35 | 40 |
| Glutamic acid | 60 | 80 | 110 | 160 |
| Asparagine | 30 | 40 | 60 | 240 |
| Glutamine | 60 | 40 | 360 | 700 |

Changes in the free amino acid composition in the seed may be indicative of developmental changes. For example, Fine and Barton (1958) showed that both the ratio of amino acids and the absolute amounts change during after-ripening of tree peony seeds.

Protein synthesis and its association with nucleic acid metabolism will be discussed in sections II5 and II6 of this chapter.

## 4. *Metabolism of Phosphorus-containing Compounds*

Phosphates play an extremely important role in a variety of reactions in seeds. Thus phosphate is required for the formation of the nucleic acids—which in turn are intimately connected with protein synthesis and the hereditary constitution of the plant cell. The function of phospholipids, such as lecithin, in controlling surface properties and permeability of cells and of its organelles is well-established today, and may be taken to be similar in the seed and seedling to that in other plant tissues. The various phosphate sugars and nucleotides are very closely linked with the energy-producing processes in the cell during germination.

Phosphorus appears in seeds primarily in the organic form and very little seems to be present as inorganic orthophosphate. Among the phosphorus-containing compounds are the nucleic acids, phospholipids, phosphate esters of sugars and nucleotides and phytin—the calcium and magnesium salt of inositol-hexaphosphoric acid. The calcium and magnesium content of the phytin molecule is variable. Thus wheat phytin contains 12 per cent calcium and 1·5 per cent magnesium, while oat phytin contains 8·3 per cent calcium and 15 per cent magnesium as well as 5·7 per cent manganese. The absolute amount of phytin is also very variable and varies not only between species but even in different varieties of the same species (Ashton and Williams, 1958). Phytin is very frequently present in many seeds and may constitute up to 80 per cent of the total phosphorus content of the seed. Some of the phosphorus-containing compounds occurring in cotton seed and the changes in them during germination are shown in Table 5.8. Because most of the phosphate is present

Table 5.8—Changes in Composition of the Various Phosphorus
Fractions in Cotton Seeds, Paymaster Variety, during Germination
The results are given as mg phosphorus per gram dry weight
(From Ergle and Guinn, 1959)

| Time of germination in days | Dry seeds | 1 | 2 | 4 | 6 |
|---|---|---|---|---|---|
| Phytin | 8·61 | 8·49 | 7·15 | 4·00 | 1·97 |
| Inorganic | 0·44 | 0·29 | 1·87 | 4·77 | 7·02 |
| Total lipid | 0·71 | 0·81 | 0·87 | 0·50 | 0·85 |
| Ester | 0·32 | 0·41 | 0·40 | 0·58 | 0·42 |
| RNA | 0·12 | 0·11 | 0·15 | 0·25 | 0·39 |
| DNA | 0·11 | 0·11 | 0·12 | 0·21 | 0·44 |
| Protein | 0·11 | 0·10 | 0·16 | 0·28 | 0·26 |

in the bound form, orthophosphate may well be a limiting factor in many of the reactions mentioned above. For this reason the large amount of phytin present may be regarded as a store of inorganic phosphate which is liberated as germination proceeds. The liberation of this phosphate is by the enzymic hydrolysis of the phytin by a phosphatase. This enzyme is known as phytase and is probably not entirely specific for phytin and is capable of hydrolysing other phosphate ester linkages as well. As will be seen in Table 5.8, the amount of phytin in cotton seed drops quite quickly during germination, so that after 6 days most of the phytin has disappeared. Concurrently, inorganic phosphate accumulates in the seeds. Similar rapid disappearance of phytin from germinating seeds has been observed in wheat, oats, peas and lettuce, as well as other seeds. In all these cases phytin is present in very considerable amounts. In cotton seeds, all the phytin is present in the cotyledons (Ergle and Guinn, 1959). It is probable that in other seeds, also, a large amount of phytin is present in the storage tissues. However, Albaum and Umbreit (1943) showed that phytin is also present in the embryo of oat, disappearing rapidly during germination, but state that the endosperm contains more phosphorus than the embryo. Some of this phosphate is transported to the embryo during germination. Usually there is a good correlation between the rapidity of phytin breakdown and phytase activity of the seed or seedling. Such correlation is demonstrated for lettuce seeds in Fig. 5.19. In oats, phytase

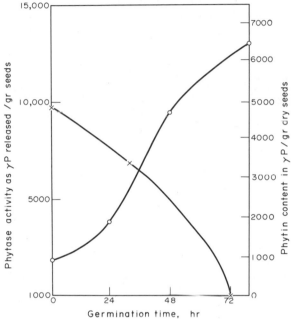

Fig. 5.19. Phytin content and phytase activity in germinating lettuce seeds.
Phytin content ×——× as γ phytin P per gram dry seeds
Phytase activity ○——○ as γP released by gram seeds
(From Mayer, 1958)

activity in the entire seeds seems to develop rather slowly and the total phytin content decreases to about a half in the course of a week. In wheat, on the other hand, most of the phytin has disappeared after a similar period. Peers (1953) showed that in the dry wheat seed about 80 per cent of the phytase is present in the endosperm and only 1 per cent in the embryo. The remainder is distributed between the other parts of the seed. The enzyme prepared from wheat was characterized by stability to fairly high temperatures. The purified enzyme was less heat resistant than the crude preparation. The optimum temperature for enzyme activity was around 51°C and the pH optimum around 5·4. The data seem to be characteristic for phytase prepared from seeds. The mode of action of phytase is quite complex and the various inositol phosphates are not attacked at equal rates.

At one time it was thought that phytin could serve as an energy source and that transphosphorylation reactions between phytin and dinucleotides could occur.

However, for thermodynamic reasons this seems very unlikely. Moreover, numerous experiments designed to show the existence of such transphosphorylation reaction were unsuccessful. Thus, although the literature still contains reports on phytin as an energy source, these must be regarded with great reserve (Mayer, 1973).

It is significant that the germinating seeds contain not only phytase but also the enzyme systems necessary for phytin synthesis. In fact some appears to be formed during germination (Lahiri Majunder *et al.*, 1972).

Little is known about the fate of the inositol which arises from phytin breakdown but it is probably normally metabolized during germination. In addition phytin breakdown releases Ca and Mg and phytin might therefore be a store not only of inorganic phosphates but also of these ions.

In addition to phytase seeds contain many phosphatases, and their activity rises during germination. These phosphatases can account for the turnover of all the phosphate esters present in the seed. An example of the changes in phosphatase and ATPase activity in germinating lettuce is shown in Fig. 5.20. In lettuce at least eight

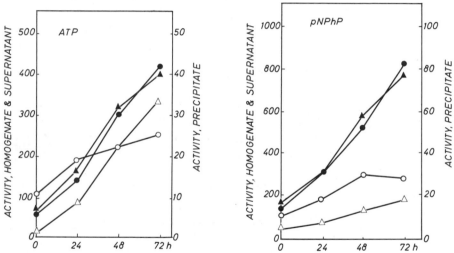

Fig. 5.20. Changes in phosphatase activity in various cell fractions during germination of lettuce. Activity as $PO_4$ released $\mu$ mol/(g dry seeds) $\times$ h for ATP and pNPhP. Homogenate ($\blacktriangle$), 2000 g precipitate ($\bigcirc$), 20,000 g precipitate ($\triangle$), 20,000 g supernatant ($\bullet$).
(From Meyer and Mayer, 1971)

distinct phosphatases were demonstrated using electrophoretic techniques. In many seeds a glycerophosphatase has been shown to be present which differs from phytase both in specificity, pH optimum and heat resistance. An active ATPase has been shown to occur in mitochondria prepared from lettuce seeds at various stages of germination. Mitochondria from a number of seeds have also been shown to have systems carrying out oxidative phosphorylation. This will be discussed in the section on respiration. Among the phosphokinases, pyruvic kinase appears to be of a very widespread occurrence. The same seeds usually contain a number of different phosphatases which are able to hydrolyse phosphoenol pyruvate.

Far fewer data are available about other phosphate containing compounds in germinating seeds. Probably this comparative lack of information is due to the difficulty in resolving the various phosphate containing compounds and analysing and estimating them accurately.

It is clear from Table 5.8 that the total changes in lipid phosphate and ester phosphate are not very large in cotton. Nevertheless, it is clear today that the phosphate-containing compounds present in membranes such as phosphatidic acid, phosphatidyl inositol, phosphatidyl ethanolamine and phosphatidyl choline are all synthetized very early during seed germination. Evidence for this is provided by the rapid incorporation of $^{32}PO_4^-$ into these fractions (Katayama and Funahashi, 1969). The rapid transformation of the phosphate containing components of the membranes is paralleled by the rather early onset of the metabolism of other membrane components. As a result, the composition and the properties of the membranes of various organelles of the seed and even of the cell membrane itself, may undergo changes during germination. This would be fully in accord with changes in the structure of subcellular organelles such as the mitchondria, glyoxysomes, ribosomes and polysomes which may be observed during seed germination (Lott and Castel-franco, 1972; Treffry *et al.*, 1967). The appearance of a better defined endoplasmatic reticulum is also characteristic of seed germination and has been observed by many workers, using the electron microscope (Srivasta and Paulson, 1968; Swift and O'Brien, 1972). In general a very large variety of sugar esters is present in seeds, including the intermediaries of glycolysis and triose phosphates, as well as nucleotides such as ATP, NAD, NADP and others such as UTP, ITP and similar and related substances. In no case has any very regular change in any of these fractions been observed. The hexose phosphate content of oat embryos rises during germination but the hexose content of the endosperm was not examined (Albaum and Umbreit, 1943). The ATP content of various parts of seedlings showed an initial rise followed by a subsequent decrease. Sebesta and Sorm (1956) showed that ATP is virtually absent in the dry bean seed and is formed during imbibition. However, after an initial rise in the ATP content the level decreases in all organs except the cotyledons. In contrast, Albaum and Umbreit (1943) showed a steady increase in the ATP content of the embryo of oats.

A detailed analysis of nucleosides and nucleotides during the first 40 hours of germination of peas was carried out by Brown (1965). His results show that the AMP level fell during this period. ADP showed an initial small drop followed by a much bigger one between 16 and 40 hours while ATP first rose for 16 hours and then fell again. Free adenosine fell markedly during germination while xanthosine doubled during the first 40 hours. A very rapid rise in the ATP content of wheat embryo has been reported. Within 1 hour the ATP content rose 10-fold (Obendorf and Marcus, 1974). The ATP content of seeds of a number of species (*Trifolium incarnatum*, *Brassica napus* and *Lolium multiflorum*) was correlated with seed vigour and was suggested as an indicator for this property (Ching, 1973).

The changes in the total NAD and NADP content of a number of seeds was studied by Bevilacqua and Scotti (1953) and Bevilacqua (1955). They showed that the nucleotide content of the seed and seedling rises in all cases during germination. The rise in peas was much greater than in wheat or oats. In both wheat and oats the initial rise and fall occur at different times in different seeds. In wheat endosperm an initial rise up to 5 days is followed by a plateau, while in peas at this time a marked drop of nucleotide content in the cotyledons is observed (Fig. 5.21).

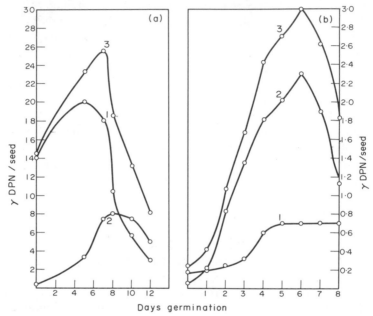

Fig. 5.21. Changes in pyridine nucleotides in seeds of *Pisum sativum* and *Triticum vulgare*
during germination.
(a)—*Pisum sativum*; (b)—*Triticum vulgare*
1—endosperm or cotyledons; 2—embryo; 3—whole seed
(Bevilacqua, 1955)

A similar change has been reported for germinating rice seeds by Mukherji *et al.* (1968), who studied the changes in each of the nucleotides, NAD, NADP, $NADH_2$ and $NADPH_2$. A massive conversion of NAD to NADP has been reported during the very onset of germination of peanuts, *Arachis* (Reed, 1970). The necessary enzyme NAD kinase rose sharply in activity at the very start of imbibition. Clearly these substances are also actively metabolized, as is consistent with their role in all energy requiring and producing processes during germination.

Of the enzymes concerned with the metabolism of nucleotides and sugar phosphates, only a few have been investigated to any extent in seeds. In general, most attention has been paid to general phosphatase activity on the one hand, and to the phosphokinases responsible for phosphate transfer, on the other hand.

The obvious and only conclusion from these results is that the nucleotides and nucleosides are rapidly and actively metabolized during germination.

## 5. *Metabolism of Nucleic Acids*

Some of the overall changes in nucleic acid content of seeds were shown in Figs. 5.6 and 5.9 and in Tables 5.1 and 5.3. The data in Fig. 5.9 and Table 5.3 appear to indicate a redistribution of nucleic acids between various parts of the seedling, mainly a decrease in their amount in the storage tissue and an increase in the growing embryo. The total nucleic acid content of the seed increases during germination as was shown for cotton, Table 5.8. Such an increase would be expected if there is an increase in cell number during germination and no storage nucleic acids are present.

During germination both cell elongation and cell division occur eventually. It is therefore obvious that at some time both protein synthesis and synthesis of nucleic acids must occur. Much attention has been given to the question—When do the various events in the normally accepted pathways of nucleic acid metabolism and protein synthesis occur in germinating seeds? There is very little evidence to indicate that the actual metabolic pathways are different in germinating seeds than in other tissues. Much of the work has been done using incorporation of radioactive precursors into the nucleic acid fractions and studies with metabolic inhibitors have been carried out. When using intact seeds many difficulties are met with, because of diffusion barriers in seed coats, uneven distribution of precursors in the seed and even differential penetration of both precursors and inhibitors into different parts of the seed. Many researchers have used isolated seed parts for such studies, but the results are often difficult to interpret and care must be taken to generalize from such investigations about the events in the whole seed.

Generally speaking DNA synthesis is detected rather late during seed germination, Table 5.8. However, in isolated wheat embryo appreciable DNA synthesis was shown to occur after 12 hours of germination. It occurs before or after cell division, which is often but not always a prerequisite of radicle protusion. In wheat embryos DNA synthesis followed prior protein synthesis (Mary *et al.*, 1972). DNA is apparently modified during germination. The properties of DNA isolated from dry seeds differ from those of the DNA extracted from germinated embryos (Chen and Osborne, 1970). The DNA of the dry seeds is very resistant to adverse conditions such as heat, or dehydration. Little is as yet known about its properties.

The synthesis of DNA is often studied by following incorporation of radioactive thymidine. This method is not always reliable. Anomalous incorporation of thymidine into cytoplasmic fractions has been reported during germination. This may well be due to the breakdown of the thymidine followed by the subsequent metabolism of the breakdown products.

There is convincing evidence that dry seeds contain long-lived stable mRNA which during germination becomes available for polysome formation. Such mRNA has been shown to occur in the ribosomal fraction of dry wheat embryos (Schultz *et al.*, 1972). In entire wheat seeds, a messenger fraction was shown to be synthesized very early, after a few hours of germination (Rejman and Buckoviz, 1973).

The amount of polysomes is usually very low in dry seeds or polysomes are entirely absent. During germination of wheat, polysomes rapidly appear in the embryos (Table 5.9). Polysome formation has been demonstrated both in electron micrographs of

Table 5.9—Ribosomes and Polysomes in Wheat Embryo during Germination
(From Marcus, 1969)
Ribosome activity determined by leucine incorporation into isolated ribosomes. Polysome content determined from adsorbance in polysome region of sucrose gradient

| Length of imbibition | Ribosome activity cpm/mg RNA | Polysome content O.D. units |
|---|---|---|
| 0 | 268 | 0·01 |
| 15 min | 6680 | 0·16 |
| 30 min | 23200 | 1·61 |
| 1·5 hr | 31900 | 2·42 |
| 6 hr | 56300 | 3·66 |

seeds at different stages of germination and by direct isolation of the ribosome and polysome fractions, using normal fractionation and ultracentrifugation techniques.

In cotton seeds ribosome formation and RNA synthesis occur during the early stages of germination (Waters and Dure, 1966). The cotton seeds also contain long-lived mRNA and the *de novo* synthesis of a protease which is dependent on the presence of a pre-existing mRNA was demonstrated (Ihle and Dure, 1969). It was further shown that the mRNA was formed during embryogenesis but translation of the mRNA is blocked at this stage, probably through hormonal control (Ihle and Dure, 1972). Thus in the cotton seed there occur RNA synthesis and ribosome formation and at the same time utilization of pre-existing mRNA for polysome formation. In some cases ribosomes must be synthesized, while in other cases sufficient ribosomes are present in the embryo for polysome formation. In castor beans the dry seed already contains some ribosomal RNA and the heavy ribosomal fraction increases very rapidly during germination (Table 5.10). It is not certain that long-lived mRNA is a

Table 5.10—Changes in RNA Content of Castor Bean Endosperm during Germination
(From Marre, 1967)
Results as $\mu$g/endosperm

| | Dry seed | Seed germinated 24 hours | Seed germinated 48 hours |
|---|---|---|---|
| Total RNA | 150 | 345 | 950 |
| Heavy ribosomal RNA | 73 | 175 | 550 |
| Light ribosomal RNA | 35 | 75 | 190 |
| Soluble RNA | 42 | 95 | 210 |

universal feature of all germinating seeds (Marre, 1967). Till now only a few species have been studied and generalizations are still not possible. Only a very few of the enzymes involved in RNA metabolism have been studied in detail. RNA polymerase activity increases prior to RNA synthesis and the enzymes required for RNA synthesis seem to be present at least in the mature seed of wheat.

The sequence of the synthesis of the various RNA fractions and the presumed existence and function of long-lived mRNA, or its absence, are of great importance in the control of protein synthesis in the germinating seed.

## 6. *Protein Synthesis and its Dependence on Nucleic Acids*

Since DNA synthesis apparently occurs fairly late during germination while that of RNA is rather rapid, the question arises: at what stage of germination does protein synthesis begin, what are the limitations to protein synthesis and what is known about the various requirements for protein synthesis and the enzymes involved in it. The onset of protein synthesis during germination has been deduced from three main kinds of evidence: (1) the appearance of an enzyme activity, or its increase during germination; (2) the failure of an activity to appear in the presence of an inhibitor of protein synthesis and (3) from studies based on the incorporation of radioactive precursors into the protein. In the most convincing experiments these techniques have been combined. Invariably some doubt can be expressed about some of the evidence regarding protein synthesis in germinating seeds. Many of the investigations have been made, for experimental convenience, with isolated parts of seeds rather than with intact tissues. Nevertheless, it is quite clear today from the various studies that protein synthesis occurs in seeds quite soon after imbibition. The exact time of synthesis seems to be rather variable. Almost invariably there is some delay in the onset of protein synthesis in the intact seed. This late onset may be caused by a number of factors, such as delay in formation of functional ribosomes, delay in monosome-polysome transition or of formation or availability of mRNA, or the unavailability of building blocks or of energy for protein synthesis. The dependence of protein synthesis in dry wheat embryo or pea cotyledons on a factor from imbibed seeds was first demonstrated by Marcus and Feeley (1964). This requirement could be replaced by polyuridylic acid (Table 5.11). Although first interpreted as demonstrating

Table 5.11—$^{14}$C Phenyl Alanine Incorporation in the Presence and Absence of Poly U by Ribosome Preparations from Peanut Cotyledons
(After Marcus and Feeley, 1964)

| | | mg ribosomal RNA added | Incorporation of amino acid cpm into protein by preparation from | | |
| --- | --- | --- | --- | --- | --- |
| | | | Dry seeds | Imbibed 1 day | Imbibed 4 days |
| No poly U | Unwashed microsomes | 0·4 | 3 | 57 | 78 |
| | | 0·8 | 6 | 90 | 109 |
| | Washed microsomes | 0·2 | 2 | 75 | 84 |
| | | 0·4 | 1 | 123 | — |
| Poly U added | Unwashed microsomes | 0·05 | 576 | 478 | — |
| | | 0·10 | 876 | 523 | 543 |

a requirement for mRNA synthesis, it is now supposed to be due to activation of masked long lived mRNA (Weeks and Marcus, 1971). Similar observations have been made on other tissues, such as isolated *Phaseolus vulgaris* axes, castor bean cotyledons, cotton, lettuce and other seeds (Mayer and Shain, 1974). Generally the technique used has been incorporation of radioactive amino acids into protein, either by ribosomal preparations or by seed tissue.

The ability of isolated aleurone tissue from *Avena fatua* to incorporate amino acids has been observed by autoradiographic methods. Within 10 minutes of imbibition, leucine H³ was incorporated into 50 per cent of the cells of the aleurone segments (Maherchandani and Naylor, 1972). The conclusion reached was that the tissue had an immediate capacity for protein synthesis, and that all the requirements for this synthesis were present in the dry mature tissue. The protein synthesis during hormone treatment of aleurone layers has been the focus of much attention. Here the *de novo* synthesis of protein in embryo-less half seeds was conclusively proved using a density gradient labelling technique in the presence of $D_2O$ (Filner and Varner, 1967). Protein synthesis during germination is therefore firmly established. From this it must not be concluded that all protein synthesis during germination and seedling formation is mediated by a preformed synthetic apparatus. On the contrary, as germination proceeds new ribosomes, polysomes and mRNA are formed. The level of enzymes required for protein synthesis increases. Aminoacyl-sRNA synthetases have been demonstrated in the cotyledons of *Aesculus* species and are evidently functional in the dry seeds (Anderson and Fowden, 1969, 1970), but their activity in germinating *Phaseolus vulgaris* increases during germination (Fig. 5.22).

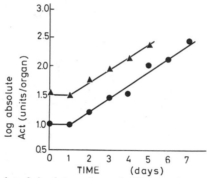

Fig. 5.22. Logarithmic plot of absolute aminoacyl-sRNA synthetase activity of plumules (●) and radicles (▲) of the french bean during germination.
(From Anderson and Fowden, 1969)

In addition to formation of active enzymes by *de novo* protein synthesis, there is good evidence for activation of preexisting inactive proteins during the early stages of germination (Marre, 1967; Mayer, 1973). Since during seed formation many enzymic systems are extremely active, it is logical to suppose that some of these became inactivated during the dehydration of the developed seeds. Many alternative ways can be suggested by which enzymes can be reversibly inactivated, such as folding, enclosure in or attachment to membranes, association or dissociation of subunits or by the addition of a section to the polypeptide chain of the active protein. In such cases during germination a reversal of these processes would occur. Thus reversible activation—inactivation would constitute a convenient and energetically economical way of storing enzymes. By such mechanisms the need for *de novo* protein synthesis would be obviated. Evidence for such activation processes is based on lack of sensitivity of development of enzyme activity to inhibitors of protein synthesis, demonstration of absence of incorporation of radioactive markers into the purified active enzyme, and if possible demonstration of the activation process in an *in vitro* system. Such complete evidence has been demonstrated in only very few cases. Probably the most complete evidence is for the formation of an amylolytic enzyme, probably amylopectin-1,6-glucosidase, in peas (Shain and Mayer, 1968a,b). In this case it was shown that while an isolated soluble or a particulate cell fraction prepared from pea cotyledons showed low enzyme activity, when these fractions were incubated together, enzyme activity doubled during incubation (Table 5.12). It was

Table 5.12—Development of Amylopectin-1,6-glucosidase Activity when Incubating Particulate and Soluble Cell Fraction from Homogenates Prepared from Dry Pea Seeds Separately or Together (After Shain and Mayer, 1968a)

| | Enzyme activity (units) | |
| --- | --- | --- |
| | Incubation time (hours) | |
| Fraction incubated | 0 | 4 |
| Particulate | 0·300 | 0·110 |
| Supernatant (soluble) | 0·990 | 1·060 |
| Particulate + supernatant | 1·290 | 2·910 |

also shown that the soluble cell fraction could be replaced by trypsin in causing activation. It seems likely that enzyme activation is of more widespread occurrence in seed germination. Probably a number of diverse mechanisms exist for activation. The relative importance of activation as opposed to *de novo* protein synthesis during the early stages of germination requires much more detailed evaluation, but data for this are not available at present.

## III. Respiration

### 1. *Gaseous Exchange*

Germination is an energy-requiring process and is therefore dependent on the respiration of the seed. In the following account we will discuss respiration both from the point of view of overall gas exchange, and from the point of view of biochemical mechanism.

In dry seeds it is almost impossible to measure either oxygen uptake or carbon dioxide output. There can be no doubt whatsoever that such gas exchange as exists in the dry seeds is at an extremely low level. The problem of measuring the gas exchange of dry seeds is further complicated by the fact that most seeds are to some extent contaminated, both on the seed coat and frequently also between the seed coat and the seed, with bacteria and fungi. These micro-organisms also have some kind of respiration and it is more than likely that some, if not all, the gas exchange measured in dry seeds is in fact due to the contaminating micro-organism. For example, Rose (1915) found that out of a hundred species examined, more than half were infected by fungi. When large bulks of seed are kept, as in grain silos, there is an appreciable heat production, resulting in a rise in temperature. The heat produced is presumably also due, at any rate in part, to the respiration of micro-organisms. It is worth pointing out that it is difficult effectively to sterilize seeds and surface sterilization may, in fact, not be sufficient. In the few established cases of very long-lived seeds, i.e. *Nelumbo*, it is difficult to understand how any form of respiration could have been maintained in the seeds for such long periods of time without entirely depleting the storage materials of the seeds, thus impairing their viability. Despite these reservations there is a certain amount of information which shows the existence of gas exchange in dry seeds. The level of gas exchange of dry seeds is very definitely dependent on the moisture content of the seeds and rises as the latter rises. Thus Bailey (1921) showed that in *Zea mays* seeds the output of carbon dioxide rose from 0·7 mg per hundred gram dry weight during 24 hours, when the seeds had a moisture content of 11 per cent, to about 60 mg when the moisture content was 18 per cent. Similar increases in carbon dioxide output with increasing moisture content have been shown for *Sorghum*, wheat and rice. The steepness of the rise in carbon dioxide output with increasing moisture content differs in different seeds.

As seeds take up water there is generally speaking a marked increase in their gas exchange. However, it has been shown that on moistening seeds with water there is an immediate gas release. This gas release seems to be a purely physical process not peculiar to seeds and involving the liberation of gas which is supposed to be colloidally absorbed within the seeds (Haber and Brassington, 1959). This observation obviously complicates any interpretation of the gas exchange of seeds immediately after they are placed in water.

Another complicating factor in respiration measurements of seeds is the presence of the seed coat. The problem of the permeability of the seed coat has already been discussed (see Chapters 3 and 4). This factor also affects the course of respiration directly. Respiration rises as the water content of the seed increases (Fig. 5.23). However, the rate of increase of respiration, as determined by $O_2$ uptake, and water uptake are by no means parallel, as can be seen from the data in Fig. 5.23. It is also clear that the increase in $O_2$ uptake consists of several phases, an initial rapid increase, a plateau when swelling has been more or less completed and a second increase. Eventually oxygen uptake of the cotyledons decreases again, probably due to senescence. In peas these different phases are observed in cotyledons of intact seeds and in cotyledons of seeds germinated without the seed coat. In the latter, oxygen uptake is a little higher (Kolloffel, 1967). However, in *Lathyrus* seeds the different phases are observed during germination of the intact seed, but are virtually absent if the testa is removed before germination (Stiles, 1935).

The changes in the oxygen uptake ($Q_{O_2}$), carbon dioxide output ($Q_{CO_2}$) and the respiratory quotient R.Q. = $Q_{CO_2}/Q_{O_2}$ of different seeds are illustrated in Figs. 5.24, 5.25 and 5.26. From these figures the general increase in both $Q_{O_2}$ and $Q_{CO_2}$ with time are evident. It will also be noted that in wheat and flax the rise of both $Q_{O_2}$ and $Q_{CO_2}$ is more or less uniform during the early stages of respiration and only falls, in the case of flax, when the seedlings become older (Fig. 5.25). In peas, on the other hand, the plateau mentioned above appears to end at about the time when the seed envelope is broken

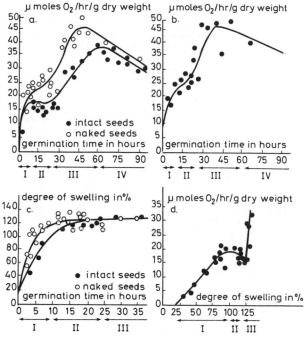

Fig. 5.23. a—The course of the respiration rate of pea cotyledons from intact germinated seeds (●) and of cotyledons from seeds germinated without seed coat (naked cotyledons ○)
b—The course of the respiration rate of excised cotyledons
c—The swelling of cotyledons from intact seeds (●) and of cotyledons from seeds germinated without seed coat (○)
d—The relation between the degree of swelling and the respiration rate of cotyledons from intact germinated seeds
(From Kolloffel, 1967)

and free gas exchange, without limitation by membranes, becomes possible (Fig. 5.24) (Spragg and Yemm, 1959). However, it is worth noting that the plateau for $Q_{O_2}$ and $Q_{CO_2}$ ends at different times. In all three cases it is important to note that although both oxygen uptake and carbon dioxide output rise with time, they rise at quite different rates. As a result the R.Q. during the early stages of germination shows very large variations. Such changes in the R.Q. are observed very often. In *Cucumis maxima* the R.Q. drops from 0·78 after 24 hours of germination to 0·41 after 72 hours. These variations point to very profound changes in the substrates used for respiration, a point which will be returned to later.

The R.Q. is dependent on the state of oxidation of the substrate oxidized. Highly oxidized substrates such as organic acids result in a R.Q. of between 1·0 and 1·5 while fats give R.Q.'s of the order of 0·7–0·8. An R.Q. of 1·0 is characteristically obtained if the substrate respired is a carbohydrate. Further, the R.Q. obtained depends on the extent to which there is genuine respiration and to what extent fermentative processes occur. In seeds with very compact tissues fermentation usually occurs initially, and only when oxygen penetrates into the tissues does respiration proper begin (Fig. 5.24).

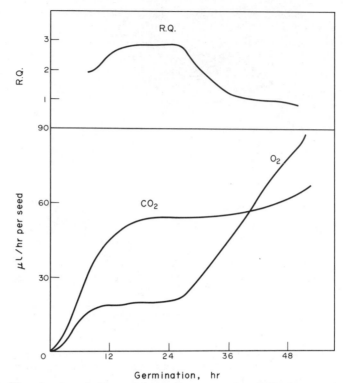

Fig. 5.24. The rate of respiration and the respiratory quotient (RQ) of pea seeds during the
first 50 hours of germination. Temp. 25°C
(Spragg and Yemm, 1959)

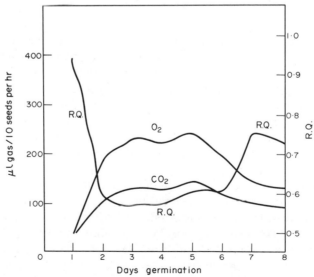

Fig. 5.25. The rate of respiration and the RQ of germinating flax seeds. Temp. 20°C
(Halvorson, 1956)

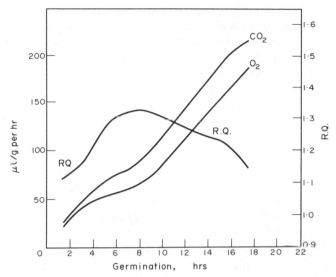

Fig. 5.26. The rate of respiration and the RQ of germinating wheat seeds. Temp. 20°C
(Levari, 1960)

In these cases there will initially be a marked carbon dioxide output due to fermentation and only a slight oxygen uptake resulting in a very high R.Q. although the substrate broken down may be a carbohydrate. In flax, Fig. 5.25, the initial R.Q. is close to one, and drops during germination as utilization of fats begins.

An indication of the difference in $Q_{O_2}$ and R.Q. in different organs of the same seed is shown in Fig. 5.27. This shows that while embryo, cotyledons and endosperm all show an initial rise in $Q_{O_2}$ the rate of oxygen uptake of the endosperm continues to rise till about 6 days and then falls again as the endosperm finally disintegrates. In contrast,

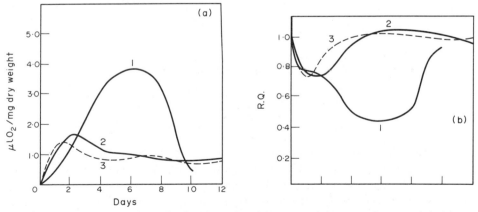

Fig. 5.27. Oxygen uptake and RQ values of various parts of the castor bean seed during
germination.
1—endosperm; 2—cotyledons; 3—embryo axis (or hypocotyl).
(a)—oxygen uptake; (b)—RQ values.
(Yamada, 1955)

cotyledons and embryo rapidly attain a steady state, if respiration is expressed per gram dry weight. These experiments were made using tissue slices of the endosperm and the hypocotyl, a method which will presumably alter the oxygen uptake of the organs as compared to their normal behaviour, both qualitatively and possibly also quantitatively. It is interesting to note that the R.Q. of the endosperm also shows very marked changes while in the embryo relatively small changes occur.

External factors which influence respiration are temperature, the oxygen and carbon dioxide content of the atmosphere and light. Each of these is liable to effect respiration somewhat differently depending on the precise stage of germination during which its influence is examined. These external factors also interact in their effect on respiration.

Generally, a rise in temperature causes an increase in the rate of respiration in the seeds. However, the early experiments of Fernandes (1923) showed that the oxygen uptake at different temperatures depends not only on the actual temperature, but also on the length of time the seeds are exposed to this temperature. In other words, in studying the effect of temperature, the time factor must also be taken into account. If raised temperatures induce dormancy, no rise in the rate of respiration is observed. The respiration of flax seeds germinated at different temperatures is shown in Fig. 5.28(a). It has also been shown that the effect of temperature depends on the presence

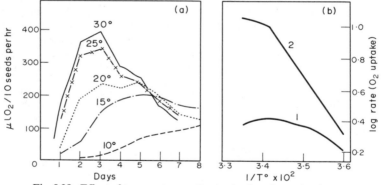

Fig. 5.28. Effect of temperature on the respiration of germinating seeds.
(a) Oxygen absorption of flax seeds during germination at different temperatures (Halvorson, 1956).
(b) The respiration of pea seeds with or without testas. Seeds were germinated for 24 hours.
1—whole seeds; 2—seeds with their testa removed.
(Spragg and Yemm, 1959)

or absence of the testa. For example, in peas whose testa had been removed, an increase in temperature raised the oxygen uptake much more than in intact pea seeds (Fig. 5.28(b)) (Spragg and Yemm, 1959). Thus temperature can only effect respiration provided oxygen can freely diffuse to the respiring tissue. If oxygen diffusion is limited, an increase in temperature will have relatively little effect.

An increase in the oxygen tension can also increase the rate of respiration of seeds. Many examples of this are available. In many cases, however, this only applies to oxygen concentrations below 20 per cent, maximum rates being reached at this value. This was shown to be the case for *Triticum spelta* and *Brassica rapa* (Reuhl, 1936). However, in *Linum usitatissimum*, in the early stages of germination when the root had just emerged, oxygen uptake had not yet reached a steady state at 20 per cent oxygen. In most cases, in the later stages of germination, when the roots were 1–3 cm long, steady states of oxygen uptake were obtained at 20 per cent oxygen. For peas it was shown that the effect of raised oxygen concentration depended on the stage of germination chosen for study. Respiration in pure oxygen was appreciably higher than in air, when the seeds had been germinated up to 36 hours. Up to this time there was a steady rise in the percentage increase of respiration in oxygen as compared to air. After 36 hours the percentage increase dropped again and at 48 hours the rates of respiration in air and in oxygen were about equal.

The dependence of respiration on the external $O_2$ concentration was also studied in *Sinapis arvense* (Edwards, 1969). As can be seen from Fig. 5.29, oxygen uptake of the

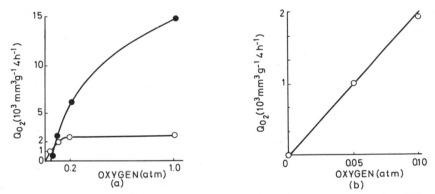

Fig. 5.29. (a) Relation of oxygen uptake ($Q_{O_2}$) and external oxygen concentration ($C_o$) of excised embryos of *Sinapis arvense* from 0 to 4 hours ○, and from 20 to 24 hours ● at 25°C; (b) enlarged scale of (a) 0 to 4 hours at 0 to 0·1 atm $O_2$.
(Edwards, 1969)

excised embryos was saturated at about 0·2 atmospheres early on in the imbibition of the embryo. However, as imbibition proceeded oxygen uptake rose as the oxygen tension increased to 1 atmosphere. From these data it appears that the seed tissues present considerable barriers to oxygen diffusion. The significance of this barrier clearly changes during germination, probably both due to changes in oxygen requirements and due to actual changes in the magnitude of the diffusion barrier.

Light has often been claimed to affect respiration of various tissues. Mostly the observed results have been obtained using white light. In view of the profound effect of light on germination previously discussed, its effect on the respiration of seeds is of some interest. It may be expected that, in so far as the total germination is increased by light, the respiration also would be increased. However, it is possible that there are other effects. The influence of red and far-red light on respiration has been studied both by Leopold and Guernsey (1954) and by Evenari, Neumann and Klein (1955). Both groups used light-sensitive lettuce seeds but their experimental techniques differed. The essential factor common to the results of both groups is that red light resulted in a raised respiration of the seeds, shortly after illumination had been applied and before any visible germination could be observed. Far-red light reversed this increase in oxygen uptake. Far-red light alone reduced the oxygen uptake. Evenari *et al.* found differences in the exact behaviour of the seeds in response to red and far-red light depending on the storage period of the seeds. Changes in the R.Q., in response to illumination, were also observed, apparently because red light raised the $Q_{CO_2}$ but did not affect the $Q_{O_2}$ while infra-red light depressed the $Q_{O_2}$ but did not affect the $Q_{CO_2}$. Leopold and Guernsey studied the effect of red light and far-red light on the respiration of other tissues known to be affected by the red and far-red mechanism. They found that there was a good correspondence between the effect of light and the response of respiration. When red light stimulated a process, stimulation of respiration resulted and a depression in respiration was observed when red light had an inhibitory effect. In every case far-red light reversed the effect of red light. A much more direct effect of red and far-red light has been demonstrated by Gordon and Surrey (1960). They were able to show that oxidative phosphorylation by mitochondria, isolated from oat coleoptiles, depended on the treatment of the intact coleoptiles with red or far-red light. However, the results were not consistent, red light sometimes depressing and sometimes stimulating oxidative phosphorylation, depending on the precise period at which it was given and the age of the coleoptiles.

The respiration of entire seeds is sensitive to externally applied respiratory inhibitors. Such inhibitors, provided they penetrate into the seeds and are not metabolized in it, act in much the same way as they act in other tissues. The relation between inhibitory action and the process of germination as a whole will be discussed later, as will the effect of other germination inhibitors as well as stimulators.

## 2. *Biochemical Aspects of Respiration*

Respiration is that process in which the substrate is oxidized through a series of steps, with the participation of oxygen as the final electron acceptor. All other processes in which oxygen does not participate are not strictly respiration. They are frequently termed 'anaerobic respiration', but the term 'fermentation' is a more correct one to use.

The mechanism of respiration and fermentation can be extremely varied. The best-established mechanism of respiration is that of glycolysis, in which the substrate is broken down to the level of pyruvate, followed by the oxidation of the pyruvate in the tricarboxylic acid or Krebs cycle. An alternative mechanism of oxidation is the direct oxidation of glucose phosphate leading to metabolism of the pentose cycle. An additional by-pass of the oxidative process is the glyoxylic acid cycle. Fermentative processes using carbohydrate as the substrate go via the glycolytic pathway. The pyruvate formed is either decarboxylated, leading to carbon dioxide formation and the resultant acetyl derivative is reduced to alcohol, or the pyruvate is directly reduced, leading to the formation of lactic acid. Although many other fermentation processes are known in micro-organisms, no information is available about their existence in plant tissues.

The main processes known with certainty to yield energy available to the organism are oxidative phosphorylation linked to the electron transport occurring in the Krebs cycle, and phosphorylation during glycolysis. The electron transport associated with the Krebs cycle is via a chain of enzymes consisting usually of dehydrogenases, flavoproteins and cytochromes, ending in cytochrome oxidase which transfers electrons directly to oxygen. Various parts of this electron-transport system are coupled to phosphorylation and the average ratio of phosphate esterified to oxygen taken up is $P/O = 3$.

Alternative electron-transport systems having different intermediaries and a different terminal oxidase are known. Among those that may function in plant tissues are the glutathione–ascorbic acid–ascorbic acid oxidase system, the phenol–phenolase system and the glycolic acid–glycolic acid oxidase systems. In none of the alternative electron transport systems mentioned is there any evidence to show coupling to phosphorylation and their precise function is unclear.

An electron transport shunt, which is cyanide resistant, is known in plant tissues. In this shunt there is one site for phosphorylation. This pathway probably also exists in germinating seeds.

We will attempt in the following to bring such evidence as is available to show the existence of these various pathways in germinating seeds and try and evaluate the relative importance of them during germination.

The existence of glycolysis may be reasonably assumed in seeds, as its presence in many plant tissues has been shown (see review by Stumpf, 1952). Conclusive evidence for its occurrence has only been brought in a few cases. Probably the clearest proof for the existence of glycolysis is for pea seeds, in which all the necessary enzymes seem to occur (Hatch and Turner, 1958). These workers prepared extracts from peas which were capable of catalysing the glycolytic reaction. The ability to carry out glycolysis was not followed through different stages of germination. However, the fact that peas can, during germination, accumulate alcohol or lactic acid or both is well established. It may therefore be supposed that during the early stages of germination, as well as in the imbibed seeds, glycolysis occurs. Hatch and Turner were unable to find any evidence for any other mechanism by which substances such as glucose or glucose phosphate were broken down and, on the contrary, in their extract there was a good correspondence between the expected formation of alcohol and carbon dioxide if only glycolysis occurred, and the actual observed values (Table 5.13). Further evidence for the existence of glycolysis has been brought by showing that extracts of both peas and

Table 5.13—Glycolytic Conversion of Various Substrates by Pea Extracts
(After Hatch and Turner, 1958)

| Substrates | $CO_2$ production moles | | Alcohol production moles | |
| | Calculated | Found | Calculated | Found |
| --- | --- | --- | --- | --- |
| Fructose 21 $\mu$moles + Fructose diphosphate 2·5 $\mu$moles | 47 | 40 | 47 | 41 |
| Glucose phosphate 25·2 $\mu$moles | 50·4 | 46 | 50·4 | 49 |
| Fructose diphosphate 25 $\mu$moles | 50 | 41 | 50 | 44 |

pea seedlings take up inorganic phosphate from the medium by glycolytic phosphory-lation (Mayer, 1959; Mayer and Mapson, 1962). This showed that the essential phosphate transferring system, linked to glycolytic reactions, functions in pea extracts.

A number of other glycolytic enzymes have also been shown in peas, for example phosphoglucomutase, phosphofructokinase (Turner and Turner, 1960) and triose isosmerase (Turner *et al.*, 1965). The former is affected by phosphate, which through its action on the fructokinase regulates the rate of glycolysis (Givan, 1972). The glycolytic system has been shown to operate at all stages of germination of *Cucurbita pepo* (Thomas and ApRees, 1972).

In a number of seeds alcohol will accumulate if they are germinated under certain conditions, such as poor aeration for lettuce (Leggatt, 1948), or high temperatures in the case of bean cotyledons (Oota *et al.*, 1956). In *Zea mays* seedlings the formation of alcohol dehydrogenase was found to be increased by anaerobic conditions, both in the scutellum and the embryonic axis. Apparently the substance causing indirect alcohol dehydrogenase formation was acetaldehyde (Hageman and Flesher, 1960). The same conditions which caused an increase in dehydrogenase were accompanied by a drop in cytochrome oxidase activity in the seedlings. Not in all of these cases has the complete glycolytic system been demonstrated, but alcohol formation and the presence of the necessary enzymes is indicative of its presence.

There is evidence for the existence of the pentose phosphate cycle in seeds. Glucose-6-phosphate and phosphogluconate dehydrogenases have been shown in a number of seeds, including wheat and lettuce. Other enzymes of the pentose phosphate cycle are present in various plant tissues. Convincing evidence for the existence of the pentose phosphate cycle have been brought for mung beans (Chakravorty and Burma, 1959). The presence of all the enzymes necessary for the oxidation of glucose-6-phosphate to ribulose phosphate and further conversion of the latter was proved (Table 5.14). Of course the existence of enzymes carrying out a

Table 5.14—Changes in the Activity of Glucose-6-phosphate and Phosphogluconate Dehydrogenase during Germination of Mung Beans (*Phaseolus radiatus*)
Results are given as units/ml crude extract for total activity and units/mg protein for specific activity
(After Chakravorty and Burma, 1959)

| Age of seedling in hours | Activity of glucose-6-phosphate dehydrogenase | | Activity of phosphogluconate dehydrogenase | |
|:---:|:---:|:---:|:---:|:---:|
| | Total | Specific | Total | Specific |
| 24 | 0·50 | 0·05 | 0·50 | 0·05 |
| 48 | 0·50 | 0·06 | 0·40 | 0·05 |
| 72 | 0·27 | 0·05 | 0·25 | 0·05 |
| 96 | 0·08 | 0·02 | 0·11 | 0·03 |

process or a reaction does not yet prove the presence of this reaction *in vivo*, but it seems very likely from other information on the metabolism of seeds that the pentose phosphate cycle may be functioning.

There is an increasing amount of evidence to indicate that the pentose cycle is particularly important during the early stages of germination and may even be involved in dormancy phenomenon (Roberts, 1972). It seems possible that one function of the pentose cycle is to provide adequate amounts of NADPH for various synthetic processes, particularly as the operation of the pentose cycle is not associated with energy production. In many cases, conditions of stress increase the percentage of the pentose phosphate cycle in the general metabolism.

The tricarboxylic acid cycle is, generally speaking, very widespread in its occurrence. Three criteria have been proposed as evidence for its functioning. These criteria demand (a) proof of the existence of the complete cycle, (b) that the cycle can be entered at any point, and (c) that any substance which is an intermediary of the cycle can be used as a substrate. According to these criteria there is little or no evidence to show that the tricarboxylic acid cycle functions in seeds. However, there are many other approaches to studying this acid cycle. Such methods include showing of the existence of various enzymes of the cycle in the tissue, its linkage to oxidative phosphorylation, oxygen uptake associated with specific substrates and so on. An alternative approach is to investigate the ability of isolated mitochondria to carry out various oxidative functions and oxidative phosphorylation. Many of these approaches have been applied to the study of respiration in seeds.

The ability of mitochondria prepared from seeds to oxidize tricarboxylic acid cycle intermediaries have been shown for peas, peanuts, mung beans, lettuce, castor beans and others. The linkage of oxidation to phosphorylation is experimentally more difficult to show. The P/O ratios obtained are usually lower than those obtained for animal material. In Table 5.15 are shown the changes in the oxidative ability of

Table 5.15— Ability of Mitochondria Isolated from Lupin Seeds to Oxidize Various
Substrates
Results as $\mu$l $O_2$/hr/mg N
(Conn and Young, 1957)

| Age of<br>Substrate        seedling | Oxygen uptake | | | |
|---|---|---|---|---|
|  | 12 hours | 2 days | 4 days | 10 days |
| Succinate | 19·0 | 70·0 | 167·0 | 208·0 |
| $\alpha$-Ketoglutarate | 49·0 | 100·0 | 122·0 | 118·0 |
| Malate | 42·0 | 46·0 | 72·0 | 72·0 |
| Citrate | 30·0 | 25·0 | 74·0 | 90·0 |
| Glutamate | 3·0 | 65·0 | 68·0 | 91·0 |
| Endogenous | 4·0 | 4·0 | 9·0 | 5·0 |

mitochondria isolated from lupin seeds and seedlings at different ages. It will be seen that the oxidative ability of the mitochondria generally rises as the age of the seedling increases but not at equal rates for different substrates. Clear proof for a cyclic process has been brought for castor beans (Neal and Beevers, 1960). If slices of castor bean endosperm were fed with labelled pyruvate, it was possible to follow the fate of the various carbon atoms of the pyruvate. Pyruvate labelled with $C^{14}$ in the one position gave rise to labelled carbon dioxide rapidly and quantitatively. When, however, the label was in the two or three positions the $C^{14}$ label spread slowly through various cell constituents and the labelled carbon was liberated only very slowly as carbon dioxide. This indicated that the C-1 carbon is removed by immediate decarboxylation while the C-2 and C-3 carbons are being cycled and only slowly take part in decarboxylation reactions. In corn mesocotyls the C-2 and C-3 carbons were not released at all as carbon dioxide. Instead they appeared in various amino acids and in protein, presumably by being diverted at various stages of the tricarboxylic acid cycle.

The behaviour of mitochondria does not necessarily reflect the behaviour of the entire cell. It is well known today that the ability of isolated mitochondria to oxidize a certain substrate depends very largely on the method of isolation. The tonicity of the medium, the presence of certain ions in the isolation medium and other preparative details all play a part in determining the activity of mitochondria, as does the exact composition of the reaction mixture in which their oxidative capacity is being tested. Further, it must be remembered that the ability of mitochondria to oxidize some of the substrates of the tricarboxylic acid cycle does not prove the existence of the cycle, because of possible by-passes of it. Additional evidence is normally necessary. Such evidence may come from tracer studies, the use of inhibitors and the chromatographic identification of the intermediaries which occur during the cyclic process. Even if the existence of a complete cycle is established, its overall contribution to metabolism of the tissue will be determined by the rate of oxidation of that substrate which is least readily utilized and is rate-limiting.

Despite the reservations which must be kept in mind with regard to work with isolated mitochondria, significant progress has recently been made by the combined study of mitochondrial ultrastructure and of mitochondrial function. Studies on the ultrastructure of lettuce and rice seeds and of cotyledons of peas, *Cucurbita maxima* and soybeans all show a quite consistent picture. As germination proceeds, the mitochondria become much better defined, their structure becomes clearer and the number of cristae in them increases (Paulson and Srivasta, 1968; Srivasta and Paulson, 1968; Ueda and Tsuji, 1971; Solomos *et al.*, 1972). This phenomenon appears to be quite general and accompanies the increasing respiratory function of the seed. The protein content of mitochondria isolated from pea cotyledons at various stages of germination increased between 3 and 18 hours of imbibition and the buoyant density changed, i.e. the position of the mitochondrial fraction in density gradient centrifugation shifted, and they became lighter (Nawa and Asahi, 1971). From this it appears that the structural changes observed in the mitochondria are accompanied by compositional changes. These compositional changes appear also to be a kind of 'maturation' during germination.

Changes in the properties of mitochondria can be caused by changes in the lipid composition of their membranes, which change their permeability. Such changes have been observed in animal mitochondria by Seligman *et al.* (1967). As the appearance and composition of the mitochondria change, so also does their function change. Mitochondria isolated from peanut embryos during imbibition showed similar changes as those of peas. In addition it was observed that the mitochondria prepared from the dry embryos were deficient in cytochrome c and do not show respiratory control. As imbibition proceeds, respiratory control and cytochrome c content become more and more normal (Wilson and Bonner, 1971). During germination, therefore, mitochondria change from an aberrant form to normal ones, which can fulfill the normal respiratory functions. In addition there is little doubt that new mitochondria are formed during germination, as are other subcellular organelles.

The increase in the functionality of entire mitochondria is also indicated by increases in individual enzymes of the Krebs cycle. Thus in peas succinate dehydrogenase, fumarate dehydratase and aconitate hydratase all increase during germination (Kolloffel and Sluys, 1970). Cytochrome oxidase and cytochrome c–NADH reductase increase quite rapidly in lettuce seeds during germination (Fig. 5.30). The increase in enzyme activity corresponds only partially to the increase in

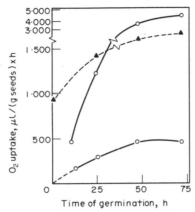

Fig. 5.30. Comparison of measured oxygen uptake of lettuce seeds during germination and potential oxygen uptake which could be mediated by activities of cytochrome c oxidase and cytochrome c reductase. Results are expressed as $O_2$ µl per hour related to 1 g seeds.
1) O ——O Upper curve. Measured oxygen uptake
2) ▲------▲ Potential oxygen uptake as calculated from cytochrome c oxidase activity
3) ●——● Lower curve. Potential oxygen uptake as calculated from NADH-cytochrome c reductase activity
(Eldan and Mayer, 1972)

water content of the seeds and it appears that the reductase undergoes an activation only partially dependent on water uptake (Fig. 5.31, cf. also Fig. 5.23). Respiratory control and ADP/O ratios change markedly in peas in the first 18 hours of germination. The respiratory control increases from 1·3 to 2·1 when malate was the substrate and from 1·3 to 3·3 with α-ketoglutarate as the substrate (Kolloffel and Sluys, 1970). Oxidative phosphorylation has been demonstrated in mitochondria from lupin seedlings with P/O ratios of over 3 when α-ketoglutarate was oxidized (Conn and Young, 1957). The exact period at which oxidative phosphorylation begins during germination is in doubt. Generally in mitochondria isolated from dry seeds oxidative phosphorylation is low or absent. Nevertheless, Wilson and Bonner (1971) were able to show some phosphorylation even in mitochondria from dry peanut embryos. In lettuce attempts to show oxidative phosphorylation in dry seeds or their mitochondria were unsuccessful (Gesundheit and Poljakoff-Mayber, 1962). However, Pradet *et al.* (1968) showed a conversion of ADP to AMP and ATP very soon after the onset of imbibition. This may indicate a limited ability to carry out oxidative phosphorylation, but might also be due to adenylate kinase reactions. Whatever the cause of the change

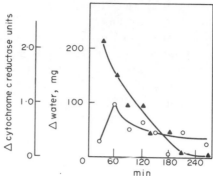

Fig. 5.31. Comparison of the rate of water uptake and the rate of development of NADH-cytochrome c reductase activity during imbibition.
Rate of change: O———O NADH-cytochrome   c reductase activity
▲———▲ Water uptake
(Eldan and Mayer, 1972)

in the relative levels of AMP, ADP and ATP during germination, the actual changed levels might have a control function according to the concept of energy change (Atkinson, 1968) as suggested by Pradet *et al.* (1968). In this respect it is interesting to note that pyruvate kinase, from cotton seeds, a key enzyme in plant respiration, has been shown to be a regulatory enzyme. Its activity is inhibited by malate citrate and ATP but stimulated by AMP and fumarate (Duggleby and Dennis, 1973). This enzyme shows some features different from pyruvate kinase isolated from other tissues.

Except for the components of the electron transport chain already mentioned, cytochrome c, cytochrome oxidase and cytochrome c–NADH reductase, information is scant. It seems probable that the other components are present, since the mitochondria can usually carry out normal oxidative and phosphorylative activity. However, direct evidence for the other components of the electron transport chain is still missing.

In addition to the conventional electron-transport system associated with the mitochondria there is scattered evidence for alternative electron-transport systems. In wheat germination it has been shown that reduced NADP can be oxidized by molecular oxygen in the presence of two enzymes, one of which is peroxidase (Conn *et al.*, 1952). This is of interest in view of the evidence previously recalled for the formation of $NADPH_2$ during the pentose phosphate shunt. The mechanism of oxidation of $NADPH_2$ and the possible energy release during its oxidation are at present in dispute. A soluble $NADH_2$-oxidase has been shown to be present in dry lettuce seeds as well as in lettuce seedlings. This latter enzyme may be a phenolase (Mayer, 1959). A complete alternative electron transport system has been shown to be present in pea seedlings. This system consists of a dehydrogenase, NADP, glutathione, and ascorbic acid in the presence of ascorbic acid oxidase. In dry pea seeds the system did not seem to function because ascorbic acid oxidase was absent (Mapson and Moustafa, 1957). This system was able to mediate 20–25 per cent of the total respiration of the young seedling and apparently was functioning in this way in the intact seedling after about 3 days of germination.

Many other oxidative enzymes are known to occur in plant tissues. Among those which have been shown to be present in seeds are catalase, peroxidase, lipoxidase as well as phenolase. The changes in these enzyme systems with germination have been followed in many cases. Unfortunately nothing is known as to how these enzymes are integrated into multi-enzyme systems and how, if at all, they are related to respiration. Phenolase has been frequently supposed to function as a terminal oxidase which can oxidize reduced co-enzymes in the presence of a phenol. There is some indication of it so functioning, for example in lettuce seeds, but this is by no means certain. Peroxidases and possibly also catalases may be connected in some way to the direct oxidation of flavoproteins by molecular oxygen, but again convincing evidence is lacking.

In summary, it can be stated that during germination of lettuce mitochondria become active. The increase in activity is accompanied by structural changes, including cristae formation. The increasing functionality is caused by an increase of all the enzyme system required for operations of the Krebs cycle. The mitochondria also develop a normally functioning electron transport chain. In the dry seed various parts of either the Krebs cycle or the electron transport chain may be deficient. The deficiency need not necessarily be the same in every tissue, nor are the exact time sequences of development of enzyme activity and functionality of the mitochondria the same in all species. Many questions are still left unanswered, and particularly whether and to what extent alternate respiratory pathways exist and play a role in seed germination. It may also be asked to what extent do respiratory pathways respond to changing environmental factors.

## Bibliography

Albaum, H. G. and Cohen, P. P. (1943) *J. Biol. Chem.* **149**, 19.

Albaum, H. G. and Umbreit, W. W. (1943) *Amer. J. Bot.* **30**, 553.

Anderson, J. W. and Fowden, L. (1969) *Plant Physiol.* **44**, 60.

Anderson, J. W. and Fowden, L. (1970) *Biochem. J.* **119**, 677.

Ashton, W. M. and Williams, P. C. (1958) *J. Sci. Food and Agr.* **9**, 505.

Atkinson, D. E. (1968) *Biochemistry* **7**, 403.

Bailey, C. F. (1921) *Univ. of Minnesota Agr. Expt. Stat. Tech. Bull.* 3.

Bevilacqua, L. R. and Scotti, R. (1953) *Acad. Ligure de Sci. et Let.* X, 1.

Bevilacqua, L. R. (1955) *Rend. Acad. Naz. Lincei ser. VII,* **18**, 214.

Boatman, S. G. and Crombie, W. M. (1958) *J. Expt. Bot.* **9**, 52.

Breidenbach, R. W. and Beevers, H. (1967) *Biochem. & Biophys. Res. Comm.* **27**, 462.

Brown, G. E. (1965) *Biochem. J.* **95**, 509.

Chakravorty, M. and Burma, D. P. (1959) *Biochem. J.* **73**, 48.

Chen, D. and Osborne, D. J. (1970) *Nature* **225**, 336.

Chibnall, A. C. (1939) *Protein Metabolism in the Plant*, Yale Univ. Press.

Ching, T. M. (1963) *Plant Physiol.* **38**, 722.

Ching, T. M. (1966) *Plant Physiol.* **41**, 1313.

Ching, T. M. (1968) *Lipids* **3**, 482.

Ching, T. M. (1973) *Plant Physiol.* **51**, 400.

Conn, E. E., Kraemer, L. M., Pei-Nan Lin and Vennesland, B. (1952) *J. Biol. Chem.* **194**, 143.

Conn, E. E. and Young, L. C. T. (1957) *J. Biol. Chem.* **226**, 23.

Cruickshank, D. H. and Isherwood, F. A. (1958) *Biochem. J.* **69**, 189.

Damodaran, M., Ramashwamy, R., Venkatesan, T. R., Mahadevan, S. and Randas, K. (1946) *Proc. Ind. Acad. Sci.* **B23**, 86.

Duggleby, R. G. and Dennis, D. T. (1973) *Arch. Biochem. & Biophys.* **155**, 270.

Duperon, R. (1958) *C.R. Acad. Sci. Paris* **246**, 298.
Duperon, R. (1960) *C.R. Acad. Sci. Paris* **251**, 260.
Dure, L. S. (1960) *Plant Physiol.* **35**, 925.
Edelman, J., Shibko, S. I. and Keys, A. J. (1959) *J. Expt. Bot.* **10**, 178.
Edwards, M. (1969) *J. Expt. Bot.* **20**, 876.
Egami, F., Ohmachi, K., Iida, K. and Taniguchi, S. (1957) *Biochimia* **22**, 122.
Eldan, M. and Mayer, A. M. (1974) *Phytochem.* **13**, 1.
Ergle, D. R. and Guinn, G. (1959) *Plant Physiol.* **34**, 476.
Evenari, M., Neumann, G. and Klein, S. (1955) *Physiol. Plant.* **8**, 33.
Fernandes, D. S. (1923) *Rec. Trav. Bot. Neerl.* **20**, 107.
Filner, P. and Varner, J. E. (1967) *Proc. Nat. Acad. Sci. U.S.* **58**, 1520.
Fine, J. M. and Barton, L. V. (1958) *Cont. Boyce Thompson Inst.* **19**, 483.
Fowden, L. (1965) In *Plant Biochemistry*, p. 361 (ed. J. Bonner and J. E. Varner), Acad. Press, New York.
Fukui, T. and Nikuni, Z. (1956) *J. Biochem. Japan* **43**, 33.
Gesundheit, Z. and Poljakoff-Mayber, A. (1962) *Bull. Res. Counc. Isr.* **D11** 25.
Givan, C. V. (1972) *Planta* **108**, 24.
Gordon, S. A. and Surrey, K. (1960) *Radiation Research* **12**, 325.
Haber, A. H. and Brassington, N. (1959) *Nature, Lond.* **183**, 619.
Hageman, R. H. and Flesher, D. (1960) *Arch. Biochem. Biophys.* **87**, 203.
Halvorson, H. (1956) *Physiol. Plant.* **9**, 412.
Hardman, E. E. and Crombie, W. M. (1958) *J. Expt. Bot.* **9**, 239.
Hatch, M. D. and Turner, J. F. (1958) *Biochem. J.* **69**, 495.
Ihle, J. N. and Dure, L. S. (1969) *Biochem. Biophys. Res. Comm.* **36**, 705.
Ihle, J. N. and Dure, L. S. (1972) *J. Biol. Chem.* **247**, 5048.
Ingle, J., Beevers, L. and Hageman, R. H. (1964) *Plant Physiol.* **39**, 735.
Ingle, J. and Hageman, R. H. (1965) *Plant Physiol.* **40**, 48.
Katayama, M. and Funahashi, S. (1969) *J. Biochem. Tokyo* **66**, 479.
Kirsop, B. H. and Pollock, J. R. A. (1957) *European Brewery Convention*, p. 84.
Klein, S. (1955) Ph.D. Thesis, Jerusalem (in Hebrew).
Kolloffel, C. (1967) *Acta Bot. Neerl.* **16**, 111.
Kolloffel, C. and Sluys, J. V. (1970) *Acta Bot. Neerl.* **19**, 503.
Kornberg, H. L. and Beevers, H. (1957) *Biochem. Biophys. Acta* **26**, 531.
Lahiri Majumder, A. N., Mandal, N. C. and Biswas, B. B. (1972) *Phytochem.* **11**, 503.
Lechevallier, D. (1960) *C.R. Acad. Sci. Paris* **250**, 2825.
Leggatt, C. W. (1948) *Canad. J. Res.* **C26**, 194.
Leopold, A. C. and Guernsey, F. (1954) *Physiol. Plant.* **7**, 30.
Levari, R. (1960) Ph.D. Thesis, Jerusalem (in Hebrew).
Lott, J. N. A. and Castelfranco, P. (1972) *Canad. J. Bot.* **48**, 2233.
McConnell, W. B. (1957) *Canad. J. Biochem. Physiol.* **35**, 1259.
McLeod, A. M. (1957) *New Phyt.* **56**, 210.
McLeod, A. M., Travis, D. C. and Wreay, D. G. (1953) *J. Inst. Brew.* **59**, 154.
Maherchandani, N. and Naylor, J. M. (1972) *Canad. J. Bot.* **50**, 305.
Mapson, L. W. and Moustafa, E. M. (1957) *Biochem. J.* **62**, 248.
Marcus, A. (1969) *Sym. Soc. Expt. Biol.* **23**, 143.
Marcus, A. and Feeley, J. (1964) *Proc. Nat. Acad. Sci. U.S.* **51**, 1075.
Marcus, A. and Velasco, J. (1960) *J. Biol. Chem.* **235**, 563.
Marre, E. (1967) *Curr. Topics Develop. Biol.* **2**, 75.
Mary, Y. Y., Chen, D. and Sarid, S. (1972) *Plant Physiol.* **49**, 20.
Marshall, J. J. (1972) *Wallerstein Lab. Comm.* **35**, 49.
Mayer, A. M. (1958) *Enzymologia* **19**, 1.
Mayer, A. M. (1959) *Enzymologia* **20**, 13.
Mayer, A. M. (1959) *Proc. Int. Bot. Cong.* vol. **2**, p. 256.
Mayer, A. M. (1973) *Seed Science & Technology* **1**, 51.
Mayer, A. M. and Mapson, L. W. (1962) *J. Exp. Bot.* **13**, 201.
Mayer, A. M. and Shain, Y. (1974) *Ann. Rev. Plant Physiol.* **25**, 167.
Meyer, H. and Mayer, A. M. (1971) *Physiol. Plant.* **24**, 95.
Mukherji, S., Dey, B. and Sircar, S. M. (1968) *Physiol. Plant.* **21**, 360.
Nawa, Y. and Ashahi, T. (1971) *Plant Physiol.* **48**, 671.
Neal, G. E., and Beevers, H. (1960) *Biochem. J.* **74**, 409.
Obendorf, R. L. and Marcus, A. (1974) *Plant Physiol.* **53**, 779.

Oota, Y. and Osawa, S. (1953) *J. Biochem. Tokyo* **40**, 649.
Oota, Y. Fujii, R. and Osawa, S. (1953) *J. Biochem. Tokyo* **40**, 649.
Oota, Y. Fujii, R. and Sunobe, Y. (1956) *Physiol. Plant.* **9**, 38.
Paech, K. (1935) *Planta* **24**, 78.
Palmiano, E. P. and Juliano, B. O. (1972) *Plant Physiol.* **49**, 751.
Paulson, R. E. and Srivasta, L. M. (1968) *Canad. J. Bot.* **46**, 1437.
Peers, F. G. (1953) *Biochem. J.* **53**, 102.
Pradet, A., Narayanan, A. and Vermeersch, J. (1968) *Bull. Soc. Franc. Physiol. Veg.* **14**, 107.
Prentice, N. (1972) *Agr. Food Chem.* **20**, 764.
Prianishnikov, D. N. (1951) *Nitrogen in the Life of Plants*, Kramer Business Service.
Pridham, J. B., Walter, M. W. and Worth, H. G. (1969) *J. Expt. Bot.* **20**, 317.
Radley, M. (1968) *Soc. Chem. Ind. London Monograph* **31**, 53.
Rebeiz, C. A., Breidenbach, R. W. and Castelfranco, P. (1965) *Plant Physiol.* **40**, 286.
Reed. J. (1970) *Dissert. Abstr. Int.* **31**, 6.
Reid, J. S. G. (1971) *Planta* **160**, 131.
Rejman, E. and Buchowicz, J. (1973) *Phytochem.* **12**, 271.
Reuhl, E. (1936) *Rec. Trav. Bot. Neerl.* **33**, 1.
Richardson, M. (1974) *Sci. Prog. Oxf.* **61**, 41.
Roberts, E. H. (1969) *Symp. Soc. Exp. Biol.* **23**, 161.
Rose, D. H. (1915) *Bot. Gaz.* **49**, 425.
Schultz, G. A., Chen, D. and Katchalski, E. (1972) *J. Mol. Biol.* **66**, 379.
Sebesta, K. and Sorm, F. (1956) *Coll. Czech. Chem. Comm.* **21**, 1047.
Seligman, A. M., Ueno, H., Morizono, Y., Wasserkrug, H. L., Katzoff, L. and Honker, J. G. (1974) *J. Histochem. & Cytochem.* **15**, 1.
Semenko, G. I. (1957) *Fiziol. Rasteny* **4**, 332.
Shain, Y. and Mayer, A. M. (1965) *Physiol. Plant.* **18**, 853.
Shain, Y. and Mayer, A. M. (1968a) *Physiol. Plant.* **21**, 765.
Shain, Y. and Mayer, A. M. (1968b) *Science* **162**, 1283.
Shain, Y. and Mayer, A. M. (1968c) *Phytochem.* **7**, 1491.
Smith, B. P. and Williams, H. H. (1951) *Arch. Biochem. Biophys.* **31**, 366.
Solomos, T., Malhotra, S. S., Prasad, S., Malhotra, S. K. and Spencer, M. (1972) *Canad. J. Biochem.* **50**, 725.
Spragg, S. P. and Yemm, E. W. (1959) *J. Expt. Bot.* **10**, 409.
Srivasta, L. M. and Paulson, R. E. (1968) *Canad. J. Bot.* **46**, 1447.
Stiles, W. (1935) *Bot. Rev.* **1**, 249.
Stumpf, P. K. (1952) *Ann. Rev. Plant Phys.* **10**, 197.
Stumpf, P. K. and Bradbeer, C. (1959) *Ann. Rev. Plant Phys.* **10**, 197.
Swain, R. R. and Dekker, E. E. (1966) *Biochim. Biophys. Acta* **122**, 87.
Swift, J. G. and O'Brien, T. P. (1972a) *Aust. J. Biol. Sci.* **25**, 9.
Swift, J. G. and O'Brien, T. P. (1972b) *Aust. J. Biol. Sci.* **25**, 469.
Tavener, R. J. A. and Laidman, D. L. (1972) *Phytochem.* **11**, 989.
Tazakawa, Y. and Hirokawa, T. (1956) *J. Biochem. Tokyo* **43**, 785.
Thomas, S. M. and ApRees, T. (1972) *Phytochem.* **11**, 2177.
Treffry, T., Klein, S. and Abrahamsen, M. (1967) *Austr. J. Biol. Sci.* **20**, 859.
Turner, D. H. and Turner, J. E. (1960) *Biochem. J.* **74**, 486.
Turner, D. H., Blanch, E. S., Gibbs, M. and Turner, S. F. (1965) *Plant Physiol.* **40**, 1146.
Ueda, K. and Tsuji, H. (1971) *Protoplasma* **73**, 203.
Varner, J. E., Ram Chandra, G. and Chrispeels, M. J. (1965) *J. Cell. Comp. Physiol.* **66** (suppl. 1) 55.
Virtanen, A. I., Berg, A. M. and Kari, S. (1953) *Acta Chem. Scand.* **7**, 1423.
Vogel, R., Trautschold, I. and Werle, E. (1968) *Proteinase Inhibitors*, Acad. Press, New York.
Waters, L. C. and Dure, L. S. (1966) *J. Mol. Biol.* **19**, 1.
Weeks, D. P. and Marcus, A. (1971) *Biochim. Biophys. Acta* **232**, 671.
Wetter, L. R. (1957) *J. Amer. Oil Chemists* **34**, 66.
Wilson, S. B. and Bonner, W. D. (1971) *Plant Physiol.* **48**, 340.
Yamada, M. (1955) *Sc. Papers Coll. Gen. Ed. Univer. Tokyo* **5**, 161.
Yamada, M. (1955) *Sc. Papers Coll. Gen. Ed. Univer. Tokyo* **7**, 97.
Yamamoto, Y. (1955) *J. Biochem. Tokyo* **42**, 763.
Yamamoto, Y. and Beevers. H. (1960) *Plant Phys.* **35**, 102.
Yocum, L. E. (1925) *J. Agron. Res.* **31**, 727.

Chapter 6

# THE EFFECT OF GERMINATION
# INHIBITORS AND STIMULATORS
# ON METABOLISM AND
# THEIR POSSIBLE REGULATORY
# ROLE

Whenever a seed becomes dormant or ceases to be dormant, something in the metabolism of the seed must change. Although this is an obvious conclusion to draw, in fact very little is known about the nature of the changes which occur. Experimentally it is inconvenient to have to wait till seeds become dormant or emerge from dormancy, as this does not permit an easy comparison of the different stages simultaneously. An alternative method of approaching the problem experimentally is to induce dormancy or to break it. For this purpose some of the compounds known to affect dormancy, such as coumarin or thiourea, can be usefully employed. Their effect on metabolism has been studied in considerable detail.

In addition gibberellic acid, cytokinins and abscisic acid (ABA) are known to affect dormancy and therefore their effect on metabolism during germination has been studied. This has thrown some light not only on the effect of these compounds themselves but also has elucidated to some extent the possible importance, in germination, of the metabolic pathways affected by them.

Although the accumulated results studying such compounds in metabolism are quite extensive, it is still difficult to obtain a really comprehensive view of what is happening during germination and dormancy breaking. It is also characteristic of this kind of work that relatively few plant species have been studied. Thus the authors have extensively studied the effect of coumarin and thiourea on lettuce seeds germination and metabolism. The action of the gibberellins have been studied primarily in the seeds of cereals such as barley or wheat and the action of cytokinins and abscisic acid have been investigated in only a few scattered species, such as cucumbers, beans and hazelnut.

In the following we will first consider the effects of gibberellins cytokinins and abscisic acid, and will then consider coumarin and thiourea. This sequence is chosen because while the former are naturally occurring plant growth substances, the latter are normally exogenously applied compounds, used as a tool to investigate metabolism during germination.

## I. Effect of Natural Growth Substances

### 1. *Gibberellic Acid*

As already mentioned (Chapter 4) gibberellins are able in some seeds to break dormancy. For this reason the metabolic effects induced by $GA_3$ are of special interest. The production by the cereal embryo of a hormone which induces hydrolysis of starch in the endosperm was first described by Kirsop and Pollock (1958). The hormone was identified as $GA_3$, independently by Yomo (1960) and by Paleg (1960).

After the initial discovery of the effect of gibberellic acid, intensive work was carried out on its effect on the metabolism of cereal seeds during germination. This work was facilitated by the early observation that the effect of $GA_3$ on the endosperm could be studied in half seeds of the cereals, from which the embryo had been removed. Such half seeds respond to $GA_3$ applied exogenously in the same way that the entire seed responds to the hormone from the embryo. The half seeds respond to $GA_3$ by the rapid hydrolysis of starch and the formation of reducing sugars, the hydrolysis of protein in the aleurone layer and the formation of amino acids, and also by the appearance of inorganic phosphate. In addition the starchy endosperm is almost completely dissolved. Some of the metabolic changes induced in the barley endosperm by $GA_3$ are illustrated in Fig. 6.1.

The data in Fig. 6.1 show that the barley seeds respond to exogenously applied $GA_3$

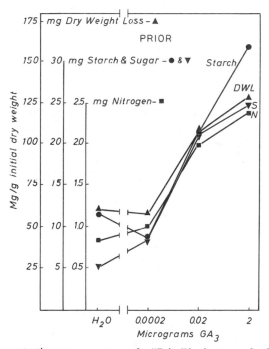

Fig. 6.1. $GA_3$ concentration-response curves for "Prior" barley treated at 30°C for 22 hours.
(Paleg, 1962)

by a release of sugar soluble N and loss of dry weight. This response was dependent on the GA₃ concentration. Attention was next focussed on the mechanism by which GA₃ brings about its effect. The work of Varner and his coworkers lead to the important discovery that gibberellic acid controlled the synthesis of α-amylase in the aleurone layer (Fig. 6.2), which led to the hydrolysis of starch in the endosperm

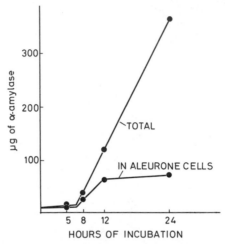

Fig. 6.2. Timecourse of α-amylase synthesis by 10 aleurone layers incubated with 1 μM GA₃. Enzyme activity was measured in the medium surrounding the aleurone layers and in the supernatant of a 0·2M NaCl extract of the aleurone layers. The term total refers to the sum of these two activities.
(Chrispeels and Varner, 1967)

(Varner *et al.*, 1965). They demonstrated that the α-amylase formed in the aleurone arose by *de novo* synthesis of protein and also showed that this protein synthesis was dependent on the synthesis of new mRNA and could be prevented by inhibitors of DNA dependent RNA synthesis. The release of sucrose from the aleurone layer was also shown to be GA₃ dependent and to require protein synthesis (Table 6.1). The effect of various factors regulating the level of α-amylase in barley seeds and probably in all cereal seeds has been summarized by Briggs (1973) (Fig. 6.3). In the scheme

Table 6.1—Dependence of Release of Sucrose from Barley Aleurone Layer on GA₃
(From Chrispeels *et al.*, 1973)

| Treatment | Duration of incubation, hours | Sucrose released μg/aleurone layer |
|---|---|---|
| Control | 10 | 120 |
| + cycloheximide | 10 | 71 |
| + GA₃ | 10 | 251 |
| + GA₃ + cycloheximide | 10 | 70 |

cycloheximide 10 μg/ml
GA₃ 2 μM

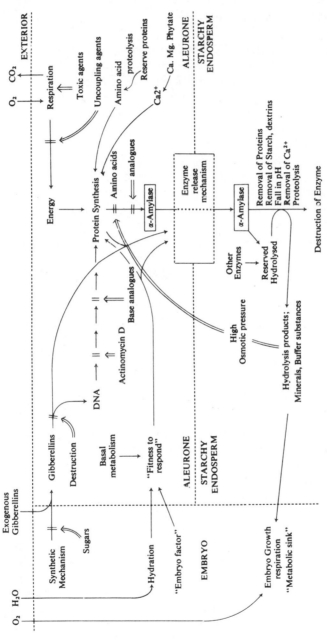

Fig. 6.3. Representation of some factors regulating the levels of $\alpha$-amylase found in whole grains of cereals. || represents inhibition. (Courtesy of D. E. Briggs, 1973 and Academic Press)

shown in Fig. 6.3 Briggs attempts to show the interrelationships between the effect of GA₃ and metabolic events occurring in the embryo, aleurone layer and endosperm. He indicates at least two possible effects of GA₃, one on $\alpha$-amylase synthesis via an affect on DNA and a second on the release of enzyme from the aleurone layer into the endosperm.

Although the effect of GA₃ on $\alpha$-amylase formation in barley endosperm is well established, it is apparently not the first response to the hormone. Pollard (1969) studied the sequential effect of GA₃ on a number of enzymes in barley and wheat half seeds. The first enzyme to be affected was a $\beta$-1,3-glucanase and this was followed by increases in activity of a phosphomonoesterase, ATPase, phytase and other enzymes. The increases in activity of the above three enzymes preceded the increase of $\alpha$-amylase and of proteolytic activity. Subsequently it was shown that very rapid responses in release of soluble sugars and phosphatases could be induced by GA₃ and that this effect could be mimicked by cyclic AMP. It has been suggested that the effect of GA₃ is mediated by cyclic AMP (Pollard, 1971). More important was the observation that at least some of the effects of GA₃, such as the initial release of soluble carbohydrates, were independent of prior *de novo* protein synthesis. Probably enzyme activation was involved. The role of gibberellic acid in the metabolism of the cereal seeds is beginning to be understood reasonably well. Nevertheless, some aspects remain quite obscure. Thus, it is known that the effect of gibberellic acid can be reversed in the cereals by the application of abscisic acid. However, ABA is not a competitive inhibitor and its mechanism is not yet fully understood. It is also known that coumarin can reverse the germination stimulating action of GA₃, for example in lettuce (Mayer, 1959), yet little is known about the mechanism of this effect.

The effect of gibberellic acid on mobilization of storage materials in the endosperm is probably not the mechanism by which germination is affected. In dormant seeds of *A. fatua*, low concentrations of GA₃, 0·1 $\mu$M, induce amylase formation but fail to break dormancy. A much higher GA₃ concentration, 0·5 mM, is required to change the dormancy markedly (Chen and Park, 1973). The changes induced by GA₃ in germination are accompanied by stimulation of the synthesis of protein and of mRNA in the embryo. Thus it appears that GA₃ has two distinct target sites, one in the embryo and one in the aleurone layer of the endosperm. The latter is not directly concerned with germination, while the former may be. It is also clear that in seeds other than the cereals, GA₃ is acting through mechanisms which do not directly involve carbohydrate metabolism. Although in seeds not belonging to the cereals some instances of induction or increase of enzyme activity by GA₃ have been reported, again the direct involvement of these effects in germination is extremely doubtful. Some studies on the effect of gibberellic acid suggest that it acts at the level of the endoplasmic reticulum (Jones, 1969) and that it may be involved in polysome formation. Other data indicate that GA₃ may affect the metabolic processes concerned with membrane formation in general (Mayer and Shain, 1974). If these findings are amplified and confirmed and it is shown that GA₃ in fact acts on membranes of cells and organelles then it may well be that its effect on seed germination may also be at this level. However, it must be stated that at present the metabolic basis of the dormancy breaking action of GA₃ remains very unclear (Jones, 1973).

## 2. *Cytokinins*

The dormancy breaking action of cytokinins has already been described in Chapter 4. The metabolic mechanisms underlying the action of the cytokinins is far from clear. During germination, possibly in its very early stages, cytokinins are converted from inactive to active forms, by mechanisms which are unknown (Van Staden, 1973). A detailed investigation of cytokinin metabolism during the germination of *Fraxinus* has been carried out using $^{14}$C labelled zeatin (Tzou *et al.*, 1973). This work shows that zeatin is accumulated in the embryo during germination and is then metabolized to the ribonucleoside and to the ribosyl mono-, di- and triphosphates. This process is quite rapid. It is interesting that the metabolism of zeatin took place in both dormant and non-dormant embryos and was not inhibited by ABA. The site of action of cytokinins in seed germination is still unclear. Expansion of the cotyledons has been suggested as the cause of stimulation of germination in lettuce (Ikuma and Thiman, 1963). However, in *Acer pseudoplantanus* cytokinins treatment results in increased elongation of the radicle (Pinfield and Stobart, 1972). The work of Ashton and his coworkers shows very clearly that the embryonic axis of squash seeds secretes cytokinins, which induce formation of isocitric lyase and proteolytic enzymes in the cotyledons (Penner and Ashton, 1967). Cytokinins released by the endosperm appear to reduce the ion release from the aleurone in wheat in some way and to affect triglyceride metabolism in the aleurone layer (Eastwood and Laidman, 1971).

It appears reasonable to assume that like GA$_3$, cytokinins are produced in some part of the seed, metabolized and transported to other parts of the seed. However, this leaves quite unanswered the question: What is the mechanism by which cytokinins react at their site of action? Work on seed germination has thrown little additional light on this question. Although effects of cytokinins on RNA metabolism are quite well documented, there is nothing to show whether this action is a primary effect or the result of other metabolic events. The mode of action of the cytokinins has recently been reviewed (Kende, 1971; Hall, 1973). At present there is no reason to assume that at the metabolic level cytokinins in germination act differently than in other developmental stages in the plant. However, this assumption remains to be proven experimentally.

## 3. *Abscisic Acid*

The function of abscisic acid in imposing dormancy on many seeds has already been described. At least in some cases the degree of dormancy correlates well with the ABA content of the embryo, e.g. in *Fraxinus americana* (Sondheimer *et al.*, 1968) as well as in other species (Milborrow, 1974). It is characteristic of exogenously applied abscisic acid that it must be continuously present. As soon as ABA is removed and the seeds are rinsed, its effect is rapidly reversed and germination can take place (Milborrow, 1974). It seems likely that this can be accounted for at least in part by the inactivation of the residual ABA in the seed. Direct evidence for inactivation of exogenously applied ABA has been provided in the case of axes of *Phaseolus vulgaris* and *Fraxinus americana.*

In *Fraxinus* two of the degradation products of $2^{14}C$ ABA were identified as phaseic acid and dehydrophaseic acid. In dormant seeds of *Fraxinus* the level of ABA dropped during stratification at 5° (Galson *et al.*, 1974). In the axis of *Phaseolus vulgaris* ABA was also degraded, but it appeared that growth of the axes was resumed when their internal ABA level was still relatively high (Walton and Sondheimer, 1972). It is also evident that ABA interacts with other growth regulators, and particularly with cytokinins and with gibberellic acid (Ketring, 1973).

Zeatin can partially reverse the effect of ABA, but does not do so by changing the rate of ABA inactivation. ABA and zeatin do not necessarily act antagonistically. For example, in *Acer pseudoplantanus* ABA did not reverse the stimulatory effect of cytokinins on radicle growth (Pinfield and Stobart, 1972).

The metabolic effect of ABA is still unclear. However, there is at least some evidence that ABA interferes in nucleic acid metabolism. Thus using techniques of autoradiography it has been shown that ABA inhibits the incorporation of $^3H$ uridine and $^3H$ thymidine into the embryos of *Fraxinus excelsior* (Villiers, 1968). Protein synthesis itself was not directly inhibited by ABA. Direct studies of DNA and RNA metabolism in pea embryos also point to an effect of ABA at the level of the nucleic acids and especially RNA, as was shown by Kahn and Heit (1968). These workers showed that ABA depressed incorporation of $^{32}P$ into various nucleic acid fractions of the embryo.

Direct evidence for ABA action on RNA metabolism has been provided. In excised cotton embryos, ABA prevented translation of preexisting mRNA which coded for known enzymes (Table 6.2). Again the effect of ABA was dependent on its continued presence. The inhibition caused by ABA was prevented when Actinomycin D, a known inhibitor of RNA synthesis, was added simultaneously (Table 6.2). Actinomycin D alone did not affect the synthesis of the enzymes studied. It appeared that the

Table 6.2—Effect of ABA on Formation of Enzyme Activity in Immature Excised Embryos of Cotton Germinated in Petri Dishes
(From data of Ihle and Dure, 1972)
(Embryo age is expressed in mg. The younger the excised embryo the smaller its weight)

| Embryo age | Days germinated | Enzyme units/cotyledon pair | |
|---|---|---|---|
| | | Carboxypeptidase | Isocitratase |
| 110 mg | 3 | 12·5 | 20·5 |
| + Actinomycin D | 3 | 12·5 | 21·1 |
| + ABA | 3 | 0 | 0 |
| + Actinomycin + ABA | 3 | 12·0 | 22 |
| 90 mg | 4 | 7·0 | 12·6 |
| + Actinomycin D | 4 | 7·1 | 13·6 |
| + ABA | 4 | 0 | 0 |
| + ABA + Actinomycin D | 4 | 7·5 | 12·5 |
| + GA₃ | 4 | 7·2 | 12·6 |
| + ABA + GA₃ | 4 | 0 | 0 |

(ABA concentration $1 \times 10^{-6}$ M. Actinomycin D concentration 20 mg/ml. GA₃ concentration $5 \times 10^{-5}$ M.)

inhibitory activity of ABA requires the continued production of some RNA, which appears to be non-ribosomal RNA (Ihle and Dure, 1972). Here, therefore, the action of ABA is exceedingly complex.

The effects of ABA were reversed both by kinetin and by $GA_3$, in the case of pea embryos. However, in the cotton embryos, $GA_3$ was quite unable to reverse the inhibition induced by ABA (Table 6.2).

Probably the effect of ABA in reversing the $GA_3$ induced formation of hydrolytic enzymes in cereal seeds can also be accounted for by its effect on nucleic acid metabolism. However, the effects observed were not simple and there was no simple stoichiometry between the effect of ABA and $GA_3$ (Jacobsen, 1973). Moreover, $GA_3$ does not reverse the ABA induced dormancy in all cases.

Although $GA_3$ and ABA do not necessarily act in a competitive manner, it does seem as if both growth substances regulate similar processes. It is worth noting that ethylene was able to partially reverse the inhibition by ABA of $\alpha$-amylase production in the barley aleurone layer (Jacobsen, 1973).

Generally the ABA content of fruits is much higher than that of seeds (Milborrow, 1974). Ihle and Dure (1972) showed that excised cotton embryos germinate readily. They make the interesting suggestion that, in cotton embryos, the function of ABA transported from the mother plant to the developing embryo is to stop enzyme formation at some stage of embryogeneis, until the seed is cut off from its supply of ABA. In their view ABA prevents vivipary of the cotton seed in the boll. Theirs is probably the most detailed study thus far of the metabolic action of ABA in germinating seeds.

In excised pea cotyledons the formation of protease activity is repressed by ABA (Yomo and Varner, 1973). This appeared to be due to repression of enzyme formation by the accumulation of amino acids. The repression was not removed by incubation with $GA_3$. In fact $GA_3$ in peas had no effect on protease activity. In contrast $\alpha$ and $\beta$-amylase activity in the excised cotyledons increased much more rapidly than in the intact seed and this was prevented by ABA. The concentration of ABA which inhibited amylase formation failed to inhibit either $O_2$ uptake or $^{14}C$ leucine incorporation (Yomo and Varner, 1973). Thus here, too, as in cotton, the effect of ABA was specific and it appears that ABA has some regulatory role.

From the above discussion it can be seen that at least in some cases, ABA acts via its effect on nucleic acid metabolism. It remains to be shown that the action of ABA in inducing dormancy and inhibiting germination can be ascribed to an effect on nucleic acid metabolism in all cases. Probably the effect of ABA on amylase production can be explained in this fashion. However, although interesting in itself, this cannot be the general explanation of dormancy induction. It seems probable that, as is the case for other growth regulators, ABA has more than one target for its action. Its interaction with other growth regulators, such as the cytokinins and gibberellins remains unexplained at the metabolic level.

The role of the naturally occurring inhibitors and stimulators should be considered not only from the point of view of their effect on some specific metabolic event. It is tempting to regard the germinating seed as a model system for investigating the hormonal activity of compounds such as $GA_3$, cytokinins and ABA. However, it seems more important to consider this kind of compound as having a regulatory role in germination as an overall process and particularly as having a regulatory effect which determines the interaction of the different parts of the same seed. From the previous discussion it is clear that gibberellin, cytokinin and ABA, and possibly IAA are not present throughout the seed at equal concentrations. Rather some of them are produced in one part of the seed, such as the embryo or embryonic axis or the aleurone layer and affect the metabolic events elsewhere, for example the starchy endosperm or the cotyledons. It is obvious that during germination a number of metabolic events occur in a carefully timed sequence and not haphazardly. The timing of such sequences is no doubt partially accounted for by nuclear control, i.e by the activation of different parts of the genome at different stages of the process of germination. Although this control need not necessarily involve additional, external, factors, their involvement seems extremely likely. The germinating seed must be able to respond, in timing its metabolic events, to changing external conditions, such as light, moisture, temperature, etc. Stimulators and inhibitors may well be involved in exercising this control over the metabolic events in the germinating seeds. Just how this control is achieved is still quite unclear.

We have seen that some of the compounds affect directly some stage of protein synthesis, either at the DNA or RNA level or at later stages. Yet not all the effects observed can be accounted for by changes in protein metabolism. Germination involves other processes, such as changes in membrane composition or permeability, enzyme activation and provision of energy. The growth substances, inhibitors and stimulators, may be involved in some of these processes. It is clear from most of the work on germination that the levels of stimulators and inhibitors are changing before and during germination. Such changes must be involved in the regulatory mechanisms. Much information is available on the effect of exogenously applied compounds, but knowledge on the mechanism of synthesis or formation of growth regulators is lacking in most cases. This is a serious gap in our understanding of their regulatory action. Moreover, although it is now clear that the growth regulator content responds to external conditions, such as stratification, almost nothing is known about how such changes are brought about. Are the growth regulators present and activated or deactivated, are they present in bound form, or must each and everyone be synthesized? In any event the question of operation of the metabolic apparatus which activates or synthesizes growth regulators must be considered.

Lately there is good reason to believe that growth regulators interact. In some cases at least the action of a stimulator is counteracted fairly specifically by an inhibitor. The general concept of hormonal control in plants suggests that a number of compounds act sequentially and that at different stages of growth and development different compounds are active, at least to varying extents. The same probably is true also for germination. The dry and the germinating seeds contain a carefully poised complement of stimulators and inhibitors. Shifts in their concentration, by whatever mechanism, are likely to be crucial in the overall regulation of germination and in the timing of the various metabolic events which occur during germination.

## II.  The Effects of Coumarin and Thiourea

These compounds will be considered jointly because they were used to stimulate induction or breaking of dormancy. Most of the work was carried out with lettuce seeds. Generally speaking, seeds were treated with coumarin to reduce germination from the 50 per cent level, in the dark, to zero and compared with others stimulated by thiourea to give 100 per cent germination. By this means the correlation between germination percentage and various metabolic events was studied.

### 1.  *Effect on Storage Materials*

Dry lettuce seeds contain a very large amount of fat and a very small amount of free fatty acids. As normal germination proceeds, the lipid content of the seeds falls after about 24 hours. In the presence of coumarin this fall is prevented entirely, while in the presence of thiourea it is delayed for almost 24 hours (Fig. 6.4(a)). Free fatty acids, on

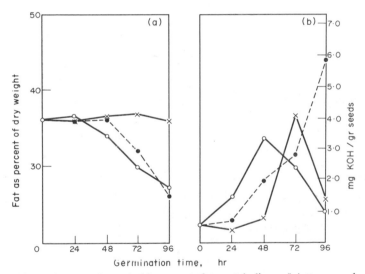

Fig. 6.4. Effect of coumarin and thiourea on fat metabolism of lettuce seeds during germination.
(Poljakoff-Mayber and Mayer, 1955)
(a) Changes in total lipids
(b) Changes in free fatty acids
O———O Seeds germinated in water
×———× Seeds germinated in coumarin (100 ppm)
●-----● Seeds germinated in thiourea (1250 ppm)

the other hand, rise continuously in the seeds treated with thiourea, while in untreated seeds and seeds treated with coumarin the free fatty acid content first rises and then falls again (Fig. 6.4(b)). At least two lipases are apparently concerned with lipid metabolism, a neutral and an acid one. The acid lipase is totally inhibited by thiourea,

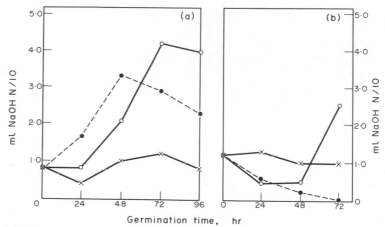

Fig. 6.5. Effect of coumarin and thiourea on lipase activity of lettuce seeds during germination (Rimon, 1957). Results as titer of alkali equivalent to fatty acids liberated by lipase.
(a) Neutral lipase; (b) Acid lipase.
(Symbols as in Fig. 6.4)

while the neutral one increases markedly in activity. Coumarin on the other hand does not affect the acid lipase *in vitro* although it prevents its rise *in vivo*. The neutral lipase is somewhat inhibited by coumarin *in vivo* (Fig. 6.5) and considerably so *in vitro*.

These changes in lipid metabolism do not suggest any primary role for lipid metabolism in dormancy and dormancy breaking. In seeds germinated in coumarin the lipids are not metabolized, because the seeds do not germinate. In seeds germinated in thiourea some of the changes are accentuated, but not altered fundamentally, during the initial steps of germination.

The changes in lipid metabolism should be considered side by side with changes in sugars, into which the lipids may be converted. In Fig. 6.6 some of these changes are illustrated. It will be seen that during germination in water or in thiourea the metabolism of glucose and sucrose is essentially the same, although the rise of glucose

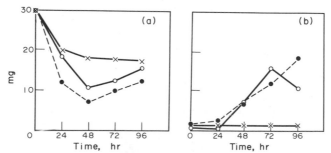

Fig. 6.6. Effect of coumarin and thiourea on sugar metabolism in lettuce seeds during germination (Poljakoff-Mayber, 1952). Results as milligram equivalents of glucose.
(a) Sucrose; (b) Reducing sugars
(Symbols as in Fig. 6.4)

in thiourea is less than in water. In seeds germinated in coumarin, again, there is no glucose formation although sucrose is broken down.

Nitrogen metabolism in germinating lettuce seeds is characterized by an increase in the amount of soluble nitrogenous compounds, the total amount of nitrogen remaining constant for the first 3 days. Between 48 and 72 hours of germination soluble nitrogen increases mainly due to seedling growth. When germination is prevented in some way, e.g. by treatment with coumarin, the rise in soluble nitrogen is prevented (Table 6.3).

Table 6.3—The Effect of Coumarin on Changes in Soluble Nitrogen
during Germination of Lettuce Seeds
Soluble N given as percentage of initial weight of seeds
(Klein, 1955)

| Time of germination in hours | Seeds germinated | | | |
| | In water | | In coumarin | |
| | % germ. | Soluble N | % germ. | Soluble N |
| 0 | | 0·260 | | 0·260 |
| 24 | 17 | 0·228 | 0 | 0·264 |
| 48 | 46 | 0·365 | 6·5 | 0·272 |
| 72 | 50 | 0·556 | 11·5 | 0·273 |

This observation suggests that during normal germination and growth, storage proteins are broken down and this breakdown is prevented by coumarin, a point of view which is strengthened by the fact that coumarin can inhibit a proteinase present in the seeds (Poljakoff-Mayber, 1953).

The proteinase inhibited by coumarin has a pH optimum of 6·8 and is present in the dry seeds. This proteinase appears to be responsible for the breakdown of an endogenous trypsin inhibitor. Removal of this inhibitor seems to be involved in permitting the formation of a trypsin-like enzyme, required for breakdown of storage proteins (Shain and Mayer, 1965). Such changes as have been observed in the storage materials all seem to be directly related to the germination process. In so far as coumarin depressed such changes, and light or thiourea stimulated them, this appears to be a direct result of a change in germination behaviour and not its cause. The same seemed to be true for the metabolism of such compounds as ascorbic acid and riboflavin in the seeds, as well as for the heavy metals which occur in various fractions of the seeds. In no case did germination inhibition or stimulation appear to affect these processes except in a secondary fashion, through their effect on germination.

However, the effect of coumarin on breakdown of proteins, via the inhibition of the proteinase might present a primary event and not simply the result of the changed germination behaviour of the seed.

## 2. *Effect on Respiration*

Some marked effects of germination stimulators and inhibitors on the respiration of seeds have been noted. These observations relate both to the effect on the gas exchange of the seeds, and also to the biochemical mechanism of respiration. Levari (1953) studied the effect of coumarin and thiourea on wheat and lettuce seeds not sensitive to light. The overall effect of coumarin on oxygen uptake and carbon dioxide output is by no means simple. At certain times coumarin raises the oxygen uptake somewhat above that of controls, but later the oxygen uptake is depressed. Another effect is the change in the respiratory quotient. Some of these changes, both for wheat and lettuce seeds, are illustrated in Fig. 6.7. A difficulty in interpreting

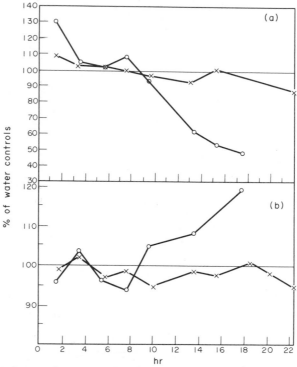

Fig. 6.7. Effect of coumarin on oxygen uptake and respiratory quotient of germinating seeds (Levari, 1953). Results expressed as percent of water controls.
(a) Oxygen uptake; (b) Respiratory quotient
○——○ Wheat seeds germinated in solutions of coumarin (250 ppm)
×——× Lettuce seeds, var. Progress (light-indifferent) germinated in coumarin (100 ppm)

precisely this kind of result lies in the fact that coumarin in these experiments completely inhibited germination. It is fairly well established that respiration rises as germination proceeds. As a result, the comparison in Fig. 6.7 is between seeds whose respiration rises because they germinate, with seeds which do not germinate at all. Klein (1955) tried to overcome this difficulty by using light-sensitive lettuce seeds. These germinate relatively little in the dark. When treated with coumarin they do not germinate at all. However, by choosing a suitable light stimulus, the effect of coumarin can be overcome and the germination restored to almost the same level as in the dark. Under these conditions the oxygen uptake of the coumarin-treated seeds is consis-

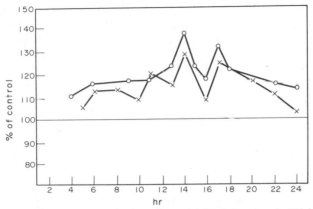

Fig. 6.8. Effect of coumarin on respiration of germinating lettuce seeds (Klein, 1955). The seeds of the light-sensitive variety Grand Rapids were germinated in 75 ppm coumarin and given a light stimulus two hours after beginning of imbibition. Germination was 7 per cent. Results as percent of water controls.

O——O Carbon dioxide output

×——× Oxygen uptake

tently higher than that of the water controls (Fig. 6.8). The carbon dioxide output behaved similarly. However, it is not clear, in this type of experiment, whether the rise in respiration was due solely to the coumarin treatment or whether the light, which increased germination, also affected respiration. Quite similar results have been obtained by Ishikawa (1958) for pea and *Setaria* seeds. He also noted that coumarin at certain concentrations raises the oxygen uptake of the seeds. He showed that this rise in oxygen uptake was quite similar to that induced in the same seeds by 2,4-dinitrophenol (DNP). Ishikawa suggests on this basis that coumarin is acting by uncoupling respiration from ATP formation in germination. This would be consistent with the fact that coumarin also increases the oxygen uptake of mitochondria isolated from lettuce seeds (Poljakoff-Mayber, 1955).

More direct evidence of the action of coumarin on oxidative phosphorylation is provided by Ulitzur and Poljakoff-Mayber (1963). *In vitro* coumarin inhibited oxidative phosphorylation and lowered the P/O ratio. In seeds germinated in coumarin oxidative phosphorylation could not be detected. Germination of seeds in thiourea advanced the stage at which oxidative phosphorylation could be detected. This stage was followed by a further rise and then a rapid drop in phosphorylative ability. Nevertheless, it was not certain whether depression or stimulation of oxidative phosphorylation could account for the changed germination behaviour of the seeds. Some of the more general observations on the effect of coumarin on metabolism during germination and growth (Mayer and Poljakoff-Mayber, 1961) could also be explained on the basis of uncoupling action, although this is probably not the entire explanation of the observed facts. In the case of barley seeds, no evidence has been found that coumarin acts as an uncoupler of oxidative phosphorylation (Van Sumere *et al.*, 1972).

A significant difference between lettuce seeds and barley is the metabolism in them of coumarin. In lettuce seeds coumarin is rapidly broken down to a number of metabolic products, which have, however, not been fully identified (Sivan *et al.*, 1965), while no evidence for metabolism of coumarin was found in barley (Van Sumere *et al.*, 1972). It is possible that the uncoupling action of coumarin is due to one of its metabolic products and not to the coumarin molecule itself. The difference in response of different species to coumarin may be related to differences in their ability to metabolize coumarin. The ability of seeds to metabolize the exogenously applied compounds is a feature which must be taken into account when using them in investigations on germination.

An important feature of the course of respiration of lettuce seeds is the rise of respiration in two stages, previously mentioned, separated by a plateau. The first rise is not necessarily related to germination as such, and occurs equally in seeds which will eventually germinate and those which will not. The rise after the plateau occurs only in those seeds which germinate and is closely associated with seedling growth. If germination is prevented in some way, the second rise in respiration is usually also prevented. If germination is prevented by elevated temperatures, which increase the initial rise in the respiratory rate, the second rise in respiration is absent.

In contrast to these marked effects of coumarin on gas exchange, virtually no effect of thiourea on either $Q_{O_2}$, or $Q_{CO_2}$ was noted for lettuce seeds, nor did the RQ change appreciably as a result of germination in thiourea (Poljakoff-Mayber and Evenari, 1958).

Various attempts have been made to study the effect of coumarin on isolated enzyme systems involved in respiration. Glycolysis seems to be operating in peas whose germination was retarded or inhibited by coumarin and therefore the glycolytic system does not appear to be affected by coumarin. Dehydrogenase activity of the tricarboxylic acid cycle of lettuce seed extracts is not affected by coumarin *in vitro*. However, the development of dehydrogenase activity *in vivo* is depressed appreciably by coumarin, presumably because the enzyme activity fails to develop (Mayer *et al.*, 1957).

Metabolism involving glucose phosphate dehydrogenase is considerably altered both by coumarin and by thiourea (Mayer *et al.*, 1966). Coumarin repressed enzyme activity and reduced its specific activity during germination. The effect of thiourea was much less marked although it, too, tended to depress total dehydrogenase activity. Since the glucose phosphate dehydrogenase is active very early in the germination process, the effect of the stimulators and inhibitors on this enzyme might be related to their action on germination itself.

Cytochrome-c oxidase activity is not affected by coumarin *in vitro*. Coumarin does appear to affect phosphate metabolism. As already mentioned earlier, the main source of phosphorus in many seeds is phytin, which releases inorganic phosphate due to its enzymic hydrolysis during germination. In lettuce seeds, phytin breakdown is completely prevented by germination in coumarin and the development of the enzyme phytase is considerably retarded. This is true even for those seeds which actually germinate in the solution of coumarin, whose phytase activity per seed was also lower than corresponding controls (Mayer, 1958). Coumarin may therefore interfere directly in the phosphate-releasing mechanism of these seeds. Moreover, in seeds germinated in coumarin, ATP failed to disappear, as it does during the first 24 hours in seeds germinated in water or thiourea. On the other hand hexose diphosphate was absent in seeds germinated in coumarin but present in seeds under the other treatments (Gesundheit and Poljakoff-Mayber, 1962).

Thiourea did not affect phytin breakdown in any way. In contrast, enzymes of the tricarboxylic acid cycle were active at a much earlier stage of development when the seeds were germinated in a solution of thiourea than in non-treated controls (Mayer, 1958). This effect was very clear despite the fact that, *in vitro*, thiourea does not stimulate any of the enzymes of the tricarboxylic acid cycle and may even depress the oxygen uptake of isolated mitochondria in the presence of some of the substrates of the cycle (Poljakoff-Mayber and Evenari, 1958).

Certain other oxidative enzymes have been shown to be affected by coumarin and by thiourea. Catalase activity of lettuce seeds, germinated in solutions of thiourea, was strongly depressed almost immediately after the onset of treatment. Coumarin only very slightly depressed catalase activity. In contrast, peroxidase activity showed a steady and persistent rise in seeds germinated in solutions of thiourea but not in the controls. It is interesting to note that light, which also promoted germination of these seeds, had quite similar effects on both catalase and peroxidase (Poljakoff-Mayber, 1953; 1956).

It has already been mentioned that other enzymes apart from cytochrome oxidase may function as terminal oxidases. Lettuce seeds contain an active phenolase which has a very high activity in dry seeds and which does not increase in amount as germination proceeds. The phenolase can also carry out the oxidation of ascorbic acid by coupled oxidation reactions. Although phenolase activity is only slightly depressed by thiourea, both *in vivo* and *in vitro*, the coupled oxidation of ascorbic acid is totally suppressed both *in vivo* and *in vitro* by thiourea. Other coupled oxidations carried out by this phenolase, e.g. of quinol, are affected by thiourea in the same manner (Mayer, 1961).

### 3. Other Effects

The action of coumarin has been investigated in relation to some other tissues and at least some of the results may be relevant to germination. Coumarin has been shown to induce swelling in isolated mesophyll cells (Harada *et al.*, 1972) and more significantly a highly specific inhibition of cellulose synthesis by coumarin has been reported (Hara *et al.*, 1973). 100 ppm coumarin inhibited incorporation of radioactive label into cellulose fractions of the cell walls of mung bean seedlings by 70 per cent. These two findings may be related and might also be of significance as far as the germination process is concerned.

A number of attempts to clarify the mechanism of action of coumarin by studying structure-activity relationships have been made. These studies attempted to determine what molecular configuration of coumarin is required for its function. It was hoped from these results to obtain an insight into the site of action of coumarin in the cell. However, these studies have by and large been disappointing and all the authors have reached the same conclusion—that no definite answer is provided by these studies (Mayer and Evenari, 1952; Berrie *et al.*, 1968; Harada and Koizumi, 1971).

Some indications have been obtained that coumarin might interfere in amino acid uptake and incorporation. This appears to be the case both in barley and in lettuce seeds as well as in yeast (Van Sumere *et al.*, 1972). However, the effect of coumarin on these events occurs rather late in germination, in contrast to the effect on the proteinase mentioned previously. Thus, it seems to be unlikely that interference in protein synthesis could account for germination inhibition, unless its effect on protein metabolism in some part of the seed is much more pronounced than that on the seed as a whole. Certainly the effect of coumarin is more pronounced in barley embryos than in entire seeds.

A different aspect of the changes in metabolism caused by germination inhibitors and stimulators is their effect on the naturally-occurring, endogenous inhibitors and stimulators. Substances active in growth and germination, present in the seeds under different treatments, have been directly extracted and assayed. The interaction of externally-applied germination stimulators with other substances has also been investigated. Lettuce seeds were extracted so as to give two fractions, an acid and a neutral one. Each of these fractions, after chromatographic separation, was assayed in coleoptile growth, lettuce seedling root elongation and lettuce seed germination tests. In the acid fraction, after 12 hours of germination in solutions of coumarin, the formation of an additional coleoptile growth inhibitor was induced and the disappearance of the natural growth inhibitors initially present was prevented. Thiourea induced the formation both of an additional coleoptile growth promoter and an inhibitor, as well as the appearance of an additional germination promoter. In the neutral fraction the only marked effect was the appearance of a germination promoting substance, 2 hours after the seeds were placed in a solution of thiourea. These changes are therefore in agreement with what might be expected from the effect of these substances, i.e. coumarin causing, in general, the appearance of new inhibitors or the failure of ones already present to disappear, while thiourea, in contrast, induced the formation of new germination promoters (Blumenthal-Goldschmidt, 1958).

Interaction studies between coumarin and thiourea in their effect on germination and also between each of them and other substances were carried out. From a study of the interaction between coumarin and thiourea it appeared that coumarin, in essence, makes seeds more sensitive to thiourea, i.e. the maximum stimulatory effect on germination becomes evident at lower thiourea concentrations. Higher thiourea concentrations, which in the absence of coumarin caused a maximum effect, are much less effective in the presence of coumarin (Table 6.4). Coumarin and thiourea interact not only in their effect on germination but also in the subsequent growth of the seedlings. The similarity between thiourea and light in stimulating germination is further reflected by the fact that coumarin makes seeds more sensitive to both agents (Mayer and Poljakoff-Mayber, 1961). An additional interaction is that between coumarin and gibberellic acid. The latter can reverse the inhibitory action of coumarin in germination. Cycosel (2-chloroethyl-trimethyl ammonium chloride) can also reverse the germination inhibition induced by coumarin. This compound also reverses the inhibition of germination by IAA (Khan and Tolbert, 1966). These results point to a possible action of coumarin on the balance of growth regulators in the seed.

Table 6.4—The Effect of Coumarin on the Germination of Lettuce Seeds in the
Presence of Thiourea
(Poljakoff-Mayber *et al.*, 1958)

| Thiourea concentration $\times 10^{-2}$ M | % germination | | | |
|---|---|---|---|---|
| | 0 | 1·0 | 2·5 | 5·0 |
| Coumarin concentration ppm | | | | |
| 0 | 33·0 | 57·8 | 59·1 | 63·4 |
| 10 | 2·7 | 15·6 | 34·3 | 14·4 |
| 20 | 0·8 | 8·5 | 13·8 | 3·6 |

In *Trifolium subterraneum* thiourea inhibits germination at low concentrations, but stimulates it at high ones (Grant-Lipp and Ballard, 1970), indicating that it acts at more than one site. Structure-activity studies on germination stimulation by thiourea have not advanced knowledge significantly. However, there is some indication that the tautomerism which thiourea and its derivatives undergo may determine its effectiveness as a germination stimulator (Mayer and Poljakoff-Mayber, 1958; Garrard and Biggs, 1966).

From studies on the uptake of thiourea by seeds, it appeared that thiourea stimulates germination while its internal concentration is still comparatively low, whereas the subsequent inhibition of seedling growth occurs when the internal concentration has risen appreciably. Thiourea shows interaction with a number of substances both in germination and growth. The kind of interaction usually differs for the two processes, possibly because of this concentration effect, as well as because of the essential difference between germination and growth. It is possible that thiourea exerts its effect by changing the level or activity of some of the growth promoters or inhibitors, but this remains to be established. Despite the fact that thiourea can complex copper and despite its effect on copper-containing enzymes, its action cannot be simply attributed to its copper-complexing action, as the effect of thiourea on germination is not simply reversed by the addition of cupric ions. The relation between thiourea and copper ions in the seeds, as reflected in the activity of copper-containing enzymes, seems to be much more complex (Mayer, 1961). Ascorbic acid, which itself does not affect germination at all, greatly enhances the stimulatory effect of thiourea. This effect was not due to interference in enzyme systems concerned with ascorbic acid metabolism but rather to some additional action of ascorbic acid (Poljakoff-Mayber and Mayer, 1961). From the results brought above it does not appear that any clear-cut conclusion can yet be drawn about the mechanism by which coumarin inhibits germination or thiourea stimulates it. This is no doubt in part due to the lack of data on normal metabolism in seeds. Nevertheless a few possible modes of action seem to emerge.

Coumarin is often supposed to function by blocking SH groups. This is primarily based on the reversal of coumarin inhibition in growth by BAL. However, Ishikawa (1958), showed that BAL does not ordinarily reverse germination inhibition caused by coumarin, and Mayer and Evenari (1952) on the basis of structure-activity studies, concluded that a blocking of SH groups could not explain coumarin action.

Coumarin may act by affecting respiratory metabolism both directly, by interfering at the phosphorylation stage, and indirectly, by affecting the availability of phosphate. It may act by preventing the formation of certain enzyme systems possibly because it prevents their liberation from some bound form. Finally, it may interfere with the formation and destruction of endogenous substances regulating germination.

Thiourea also acts by affecting the respiratory mechanism, possibly by rapidly channelling all respiration in the direction of energy yielding processes. A further way by which thiourea may act is by changing the nature and amount of the germination regulators present in the seeds.

The effect of coumarin and thiourea on metabolism of germinating seeds has been dealt with at some length because these substances have rather specific effects in seeds during germination. Such investigations serve to indicate some of the problems and achievements in metabolic studies during germination even though they do not solve them unequivocally.

## III.  The Effect of Various Metabolic Inhibitors

### 1.  *DNP*

Other substances which have more general effects on metabolism can inhibit germination. The effect of these substances on the metabolism of germinating seeds has been studied to some extent in order to determine whether their action here is the same as in metabolism in general. Special attention has been paid to the effect of 2,4-dinitrophenol (DNP). The effect of DNP on the germination of pea and lettuce seeds is shown in Fig. 6.9. Lettuce seeds are apparently much more sensitive to DNP than are peas if the concentration giving 50 per cent inhibition is considered. The inhibition of germination by DNP, in lettuce, was not reversible by light. DNP up to concentrations of $10^{-3}$ M did not effect either glucose or sucrose metabolism in lettuce in any way. The metabolism of nitrogenous compounds was somewhat affected by inhibitory concentrations of DNP, but the effects were not clear-cut, and did not account for inhibition of germination. At low concentrations DNP raises the oxygen uptake of seeds in a fashion entirely similar to that observed in other tissues, oxygen uptake was increased by about 30 per cent by a DNP concentration of $5 \times 10^{-4}$ M and the carbon dioxide output was raised about 50 per cent above the controls (Klein, 1955). In peas, oxygen uptake was increased about 10–15 per cent by a DNP concentration of $10^{-4}$ M (Ishikawa, 1958). In both cases these concentrations coincided reasonably well with those which inhibited germination. It seems, therefore, entirely reasonable to assume that DNP inhibits germination because it acts as an uncoupling agent which interferes with the energy supply of the germinating seeds.

An unusual effect of DNP has been observed in *Trifolium subterraneum*. In dormant seeds of this species 0·1 mM DNP induced germination (Ballard and Grant-Lipp, 1967). According to these workers other uncoupling agents have a similar effect on

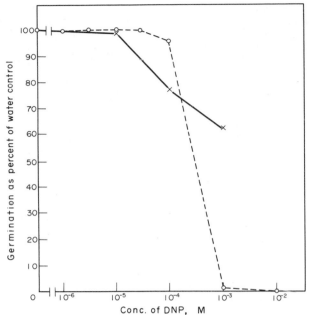

Fig. 6.9. Effect of various concentrations of dinitrophenol (DNP) on germination of lettuce and pea seeds. Results as percent of germination of the water controls. (Compiled from the results of Poljakoff-Mayber, unpublished; Ishikawa, 1958)

    O———O Lettuce
    ×———× Peas

dormancy. However, no similar effects are known in other seeds and *Trifolium* is in some respects an unusual seed. Roberts (1969) has advanced an interesting hypothesis on dormancy breakage, which is induced by many respiratory inhibitors. According to this hypothesis the pentose phosphate pathway may be restricted in dormant seeds and respiratory inhibitors break dormancy by stimulating this pathway. At present this still seems a rather speculative suggestion, although at least some supporting evidence is available.

### 2. *Other Metabolic Inhibitors*

Other respiratory inhibitors which have been studied are cyanide, azide, hydroxylamine, ethionine, diethyldithiocarbamate, iodoacetate and p-nitrophenol. The effect of some of these substances on germination is shown in Fig. 6.10.

New insight into the action of cyanide at low concentrations emerges from the work of Taylorson and Hendricks (1973). These workers showed that at low concentrations, 0·1–1·0 mM, cyanide can promote the germination of lettuce, *Amaranthus* and *Lepidium*, confirming earlier reports of the germination promoting action of cyanide. Cyanide in these species is rapidly metabolized. In the presence of cysteine it is converted to cyanoalanine, which in turn is converted to asparagine. The products of cyanide metabolism are incorporated into seed proteins during germination as was

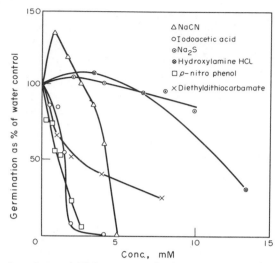

Fig. 6.10. Effect of respiratory inhibitors on germination of lettuce seeds (Mayer, Poljakoff-Mayber and Appleman, 1957). The seeds used were of the light-sensitive variety Grand Rapids. Tests carried out at 26°C in the dark.

demonstrated by the use of $^{14}CN$. The stimulating concentrations of cyanide had only a marginal effect on $O_2$ consumption. According to Taylorson and Hendricks the incorporation of cyanide into protein via asparagine and aspartic acid is adequate to explain the promotion of germination. Such an interpretation requires that a specific protein is limiting germination and that it specifically requires prior aspartic acid formation. Such an interpretation seems rather unlikely. Nevertheless these results are of importance because they show not only the metabolism of a potentially toxic compound by germinating seeds, but the actual incorporation of part of its molecule into seed proteins.

Ethionine at $5 \times 10^{-3}$ M and azide at $10^{-3}$ M also inhibited the germination of lettuce. The effect of ethionine was partially reversed by light but that of azide was not (Mancinelli, 1958).

When the effect of some of these substances on the respiration of seeds was studied, it was found that the response differed when whole seeds were treated from that obtained when homogenates were used. There was some indication that the permeability of whole seeds was a factor in the response obtained. Different inhibitors penetrated the seeds at different rates, some very quickly. Among those which penetrated rapidly were cyanide and diethyldithiocarbamate. When the inhibitors were added to homogenates prepared from seeds germinated in water the effects were observed immediately. However, the effect obtained differed, depending on the period for which the seeds had been germinated (Table 6.5). Ethionine and azide, at concentrations which inhibit germination, depressed oxygen uptake by 20–50 per cent and also changed the RQ of the seeds (Mancinelli, 1958).

Again it appears that the germination-inhibiting effect of these compounds is consistent with their effect on respiration, which is more general and applies to many

Table 6.5—Relative Oxygen Uptake of Whole Seeds and Seed Homogenates in the Presence of Various Respiration Inhibitors Added Either *in vivo* or *in vitro*. (The oxygen uptake of water germinated seeds or homogenates prepared from them is taken as 100. Oxygen uptake measured after 60 minutes) (Mayer, Poljakoff-Mayber and Appleman, 1957)

| Inhibitor added during germination or to Warburg vessels | Concentration mM | Oxygen uptake of whole seeds | | Oxygen uptake of homogenates of seeds | |
|---|---|---|---|---|---|
| | | Germinated in water, inhibitor added to Warburg vessel | Germinated in inhibitor | Germinated in water, and inhibitor added to homogenate | |
| | | 18 hour | 18 hour | 18 hour | 48 hour |
| (1) | (2) | (3) | (4) | (5) | (6) |
| Sodium cyanide | 0·8 | 82 | 125 | 47 | 41 |
| Diethyldithio-carbamate | 1·0 | 86 | 108 | 127 | 114 |
| Hydroxylamine-hydrochloride | 10·0 | 114 | 66 | 0 | 26 |
| Iodoacetate | 2·0 | 105 | 22 | 90 | 95 |
| Sodium sulphide | 10·0 | 100 | 70 | 60 | 58 |
| p-Nitrophenol | 1·0 | 100 | 87 | 105 | 76 |

other tissues. In so far as the response in seeds and seedlings differed, this can be ascribed to the fact that not all the respiratory mechanisms are functioning equally at different stages of germination. Moreover, it appears that at least some of these compounds can be broken down by the seeds, for example diethyldithiocarbamate.

Penicillin and streptomycin can also inhibit germination (Mancinelli, 1958) and respiration was inhibited at the same time (Table 6.6). However, at low concentrations

Table 6.6—Effect of Some Metabolic Inhibitors on Respiration of Lettuce Seeds Respiration as $\mu$l gas exchanged. Seeds germinated for 16 hours at 22°C (Compiled from Mancinelli, 1958)

| | Water | Ethionine $5 \times 10^{-2}$ M | Sodium azide $10^{-3}$ M | Penicillin $3·5 \times 10^{-3}$ M | Streptomycin $8·5 \times 10^{-4}$ M |
|---|---|---|---|---|---|
| $Q_{O_2}$ ($\mu$l) | 66 | 43 | 32 | 37 | 31 |
| $Q_{O_2}$ % control | 100 | 65 | 48 | 56 | 47 |
| $Q_{CO_2}$ ($\mu$l) | 56 | 34 | 39 | 28 | 24 |
| $Q_{CO_2}$ % control | 100 | 60 | 62 | 58 | 43 |
| RQ | 0·85 | 0·79 | 1·10 | 0·76 | 0·77 |
| RQ % control | 100 | 93 | 129 | 89 | 90 |

these substances could stimulate germination, an effect which was apparently reversed both by light and by higher temperatures. From interaction studies with light and temperature Mancinelli concluded that the germination-inhibitory effect of streptomycin could be accounted for entirely by its inhibition of respiration, but that this was not true for the action of penicillin. No further metabolic studies were made so that these conclusions must at present be regarded, at best, as tentative. Other fungal metabolites which inhibit germination are ramulosin and aflatoxin. Their precise mode of action is unknown.

## 3. *Herbicides*

An entirely different type of compounds affecting germination is that of the herbicides. These do not in fact present any homogeneous group since many widely different substances are used as herbicides. Many of these substances when applied directly to seeds will prevent their germination. In agricultural practice herbicides are not normally used to prevent germination. They are, however, frequently used to kill the freshly-emerging weed seedling, while leaving untouched or less affected the deeper sown crop seeds. In this case their selectivity is chiefly based on depth of penetration into the soil, rather than on any inherent physiological differences between the species to be eradicated and that to be left. Very little is known regarding the effect of herbicides on the metabolism of seeds. Their effect on metabolism in general has also not been studied very extensively, except for a few cases. The compound most widely investigated has been 2,4-dichlorophenoxy acetic acid (2,4-D). This substance has been taken to be representative of all auxin-type herbicides and has served as a model in many reactions in which the mode of action of indolyl acetic acid was being investigated. Generally speaking, 2,4-D induces profound changes in the metabolism of treated plants. These occur in the metabolism of nitrogenous compounds, carbohydrates and to some extent respiration. Thus, 2,4-D has been found to cause a more rapid use of carbohydrates, to interfere in the normal regulating action of endogenous growth substances, possibly in some case by increasing the amount of unsaturated lactones of the coumarin type such as scopoletin and methyl umbelliferone in the tissue, and finally to change the ratio between soluble and non-soluble nitrogenous compounds, by increasing the former. Scattered evidence points to 2,4-D increasing inorganic phosphate, due to decreases in the amount of phosphorylated sugars. Many enzymes have been shown to be affected to a greater or lesser extent by 2,4-D. These effects have not, however, been such as to account for the herbicidal action of 2,4-D. In fact, so far no single unambiguous effect of 2,4-D on metabolism has been observed. The literature on the action of herbicides is huge and cannot be reviewed here, particularly as the number of active compounds is constantly increasing (Audus, 1964; Moreland, 1967).

In germinating lettuce and wheat no changes in carbohydrate metabolism were induced by 2,4-D during the first 24 hours of germination (Levari, 1960). It is of course possible that at a later stage effects similar to those noted in other tissues would result.

Respiration of wheat and lettuce was increased somewhat by 2,4-D, and this increase was accompanied by changes in the respiratory quotient. Further evidence for the effect of 2,4-D on respiration comes from experiments with isolated mitochondria. In these, increasing 2,4-D concentrations progressively inhibited oxygen uptake and phosphate uptake and lowered the P/O ratio (Fig. 6.11). 2,4-D was also found to change the carbohydrate metabolism of root tips of seedlings of a number of plants. In such root tips, 2,4-D when applied to intact seedlings, depressed glycolysis and increased metabolism via the pentose cycle. If applied to seedlings

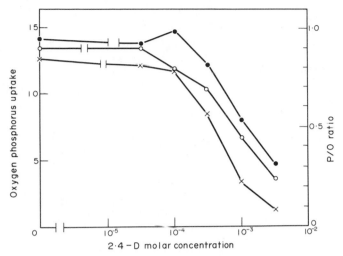

Fig. 6.11. Effect of 2,4-D on oxygen uptake and phosphorylation by soybean mitochondria,
with succinate as substrate.
(Compiled from data of Switzer, 1957).
○———○ Oxygen uptake
×———× Phosphorus uptake
●———● P/O ratio

from which the cotyledons had been removed, glycolysis was not depressed but metabolism through the pentose cycle was nevertheless increased (Humphreys and Dugger, 1957). This has been further supported by the findings that 2,4-D treatment actually increases the amount of enzymes which participate in the pentose cycle (Black and Humphreys, 1960).

Morphactins have been shown to inhibit germination. They are known to interact with many other plant growth regulators (Schneider, 1970).

From the above discussion it appears that many changes in metabolism are evoked by substances which *inter alia* also inhibit germination. Of the substances discussed, only for DNP and possibly for streptomycin, can their germination-inhibiting action be ascribed to changes in phosphorylation and respiration respectively. In all the other cases there is no possibility of relating their effects on germination to any definite metabolic process. There is no reason at present to assume that germination can be inhibited by only one kind of mechanism. It is not evident even that any given germination inhibitor acts on a single metabolic process. On the contrary, it appears that many compounds which inhibit germination do so simply because they interfere in a general way in normal metabolism. It is, however, possible that specific germination inhibitors act on some single stage of metabolism. Too few data are at present available to reach any definite conclusion on this point.

# Bibliography

Audus, L. J. (ed.) (1964) *The physiology and biochemistry of herbicides.* Acad. Press London.
Ballard, L. A. T. and Grant Lipp, A. E. (1967) *Science* **156**, 398.
Berrie, A. M. M., Parker, W., Knights, B. A. and Hendrie, M. R. (1968) *Phytochem.* **7**, 567.
Black, C. C. and Humphreys, T. E. (1960) *Plant Physiol.* **35**, XXVII.
Blumenthal-Goldschmidt, S. (1958) Ph.D. Thesis. Jerusalem (in Hebrew).
Briggs, D. E. (1973) In *Biosynthesis and Its Control in Plants*, p. 219 (ed. Milborrow), Acad. Press, New York.
Chen, S. C. and Park, W-M. (1973) *Plant Physiol.* **52**, 174.
Chrispeels, M. J. and Varner, J. E. (1967) *Plant Physiol.* **42**, 398.
Chrispeels, M. J., Tenner, A. J. and Johnson, K. D. (1973) *Planta* **113**, 35.
Eastwood, D. and Laidman, D. L. (1971) *Phytochem.* **10**, 1459.
Galson, E., Tinelli, E. and Walton, D. (1974) *Plant Physiol.* **54**, 803.
Garrard, L. A. and Biggs, R. H. (1966) *Phytochem.* **5**, 103.
Gesundheit, Z. and Poljakoff-Mayber, A. (1962) *Bull. Res. Counc. Isr.* **11D**, 25.
Grant Lipp, A. E. and Ballard, L. A. T. (1970) *Z. Pflanzenphysiol.* **62**, 83.
Hall, R. H. (1973) *Ann. Rev. Plant Physiol.* **24**, 415.
Hara, M., Umetsu, N., Miyamoto, C. and Tamori, K. (1973) *Plant & Cell Physiol.* **14**, 11.
Harada, H. and Koizumi, T. (1971) *Z. Pflanzenphysiol.* **64**, 350.
Harada, H., Ohyama, K. and Cheruel, J. (1972) *Z. f. Planzenphysiol.* **66**, 307.
Humphreys, T. E. and Dugger, W. M. (1957) *Plant Physiol.* **32**, 136.
Ihle, J. N. and Dure, L. S. (1972) *J. Biol. Chem.* **247**, 5048.
Ikuma, H. and Thiman, K. V. (1963) *Plant & Cell Physiol.* **4**, 113.
Ishikawa, S. (1958) Kummamoto, *J. Science Ser.* **B4**, 9.
Jacobsen, J. V. (1973) *Plant Physiol.* **51**, 198.
Jones, R. L. (1969) *Planta* **87**, 119.
Jones, R. L. (1973) *Ann. Rev. Plant Physiol.* **24**, 571.
Kende, H. (1971) *Inter. Rev. Cytol.* **31**, 301.
Ketring, D. L. (1973) *Seed Sci. & Technol.* **1**, 305.
Khan, A. A. and Heit, E. E. (1969) *Biochem. J.* **113**, 707.
Khan, A. A. and Tolbert, N. E. (1966) *Physiol. Plant.* **19**, 76.
Kirsop, B. H. and Pollock, J. R. A. (1958) *J. Inst. Brew.* **64**, 22.
Klein, S. (1955) Ph.D. Thesis, Jerusalem (in Hebrew).
Levari, R. (1953) *Pal. Bot. Jer. Ser.* **6**, 47.
Levari, R. (1960) Ph.D. Thesis, Jerusalem (in Hebrew).
Mancinelli, A. (1958) *Ann. di Botanica* **26**, 56.
Mancinelli, A. (1958) *Ann. di Botanica* **26**, 67.
Mayer, A. M. (1958) *Enzymologia* **19**, 1.
Mayer, A. M. (1959) *Nature* **184**, 826.
Mayer, A. M. (1961) *Physiol. Plant.* **14**, 322.
Mayer, A. M. and Evenari, M. (1952) *J. Exp. Bot.* **3**, 246.
Mayer, A. M. and Poljakoff-Mayber, A. (1958) *Bull. Res. Council Israel* **6D**, 103.
Mayer, A. M. and Poljakoff-Mayber, A. (1961) In *Plant Growth Regulation*, p. 735, Iowa State University Press.
Mayer, A. M., Poljakoff-Mayber, A. and Appelman, W. (1957) *Physiol. Plant.* **10**, 1.
Mayer, A. M., Poljakoff-Mayber, A. and Krishmaro, N. (1966) *Plant & Cell Physiol.* **7**, 25.
Mayer, A. M. and Shain, Y. (1974) *Ann. Rev. Plant Physiol.* **25**, 167.
Milborrow, B. V. (1974) *Ann. Rev. Plant Physiol.* **25**, 259.
Moreland, D. E. (1967) *Ann. Rev. Plant Physiol.* **18**, 365.
Paleg, L. G. (1960) *Plant Physiol.* **35**, 293.
Paleg, L. G. (1960) *Plant Physiol.* **35**, 902.
Paleg, L. G. (1962) *Plant Physiol.* **37**, 798.
Penner, D. and Ashton, F. M. (1967) *Plant Physiol.* **42**, 791.
Pinfield, N. J. and Stobard, A. K. (1972) *Planta* **104**, 134.
Poljakoff-Mayber, A. (1952) *Pal. J. Bot. Jer. Ser.* **5**, 186.
Poljakoff-Mayber, A. (1952) *Bull. Res. Council, Israel* **2**, 239.
Poljakoff-Mayber, A. (1953) *Enzymologia* **16**, 122.
Poljakoff-Mayber, A. (1953) *Pal. J. Bot. Jer. Ser.* **6**, 101.

Poljakoff-Mayber, A. (1953) *J. Exp. Bot.* **6**, 313.
Poljakoff-Mayber, A. and Evenari, M. (1958) *Physiol. Plant.* **11**, 84.
Poljakoff-Mayber, A. and Mayer, A. M. (1955) *J. Exp. Bot.* **6**, 287.
Poljakoff-Mayber, A. and Mayer, A. M. (1961) *Ind. J. Plant Physiol.* **3**, 125.
Poljakoff-Mayber, A., Mayer, A. M. and Zacks, S. (1958) *Bull. Res. Council, Israel* **6D**, 117.
Pollard, C. J. (1969) *Plant Physiol.* **44**, 1227.
Pollard, C. J. (1971) *Biochim. Biophys. Acta* **252**, 553.
Rimon, D. (1957) *Bull. Res. Council, Israel* **6D**, 53.
Roberts, E. H. (1969) *Symp. Soc. Exp. Biol.* **23**, 167.
Schneider, G. (1970) *Ann. Rev. Plant Physiol.* **21**, 499.
Shain, Y. and Mayer, A. M. (1965) *Physiol. Plant.* **18**, 853.
Sivan, A., Mayer, A. M. and Poljakoff-Mayber, A. (1965) *Isr. J. Bot.* **14**, 69.
Sondheimer, E., Tzou, D. S. and Galson, E. C. (1968) *Plant Physiol.* **43**, 1443.
Switzer, C. M. (1957) *Plant Physiol.* **32**, 42.
Taylorson, R. B. and Hendricks, S. B. (1973) *Plant Physiol.* **52**, 23.
Tzou, D. S., Galson, E. C. and Sondheimer, E. (1973) *Plant Physiol.* **51**, 894.
Ulitzur, S. and Poljakoff Mayber, A. (1963) *J. Exp. Bot.* **14**, 95.
Varner, J. E., Ram Chandra, G. and Chrispeels, M. J. (1965) *J. Cell. Comp. Physiol.* **66**, suppl. 1, pp. 55–68.
Van Staden, J. (1973) *Physiol. Plant.* **28**, 222.
Van Sumere, C. F., Cottenie, J., de Greef, J. and Kint, J. (1972) *Rec. Adv. Phytochemistry* **4**, 165.
Villiers, T. A. (1968) *Planta* **82**, 342.
Walton, D. C. and Sondheimer, E. (1972) *Plant Physiol.* **49**, 285.
Yomo, H. (1962) Hakko Kyokaishi **18**, 603.
Yomo, H. and Varner, J. E. (1973) *Plant Physiol.* **51**, 708.

Chapter 7

# THE ECOLOGY OF GERMINATION

In previous chapters the effect of various factors on germination of seeds has been considered. In the following an attempt will be made to relate these observations to the behaviour of seeds in their natural habitat. Various mechanisms regulate the germination of seeds in their natural habitat, some of which are internal, whereas others are external environmental factors, any of which can determine whether a given seed will germinate in a certain place or not. However, the demonstration in the laboratory of the existence of a regulating mechanism is not proof of its operation under natural conditions. To prove this and to show that the plant derives some definite advantage from a regulatory mechanism is much more difficult. The only advantage about which one may justifiably speak is where some special mode of germination has survival value for the species. This implies that the existence of the mechanism under study enables a species to exist under a given set of conditions, while the absence of this mechanism will prevent it from surviving. Survival value may also take the form of giving a species a better chance to establish itself in competition with other species occupying the same habitat. Unfortunately the proof of survival value of germination-regulating mechanisms is not easily obtainable and relatively few detailed studies have been made.

A spread of germination over a period of time can have survival value. This can protect the species from eradication as a result of adverse conditions which follow germination. If all seeds germinated simultaneously, under such adverse conditions, and the plants are unable to complete their life cycle, then no further renewal of the species would be possible.

The spread of germination over a period of time has been observed in many species (Koller, 1972). The underlying mechanisms have in most cases not been fully elucidated. Undoubtedly, such mechanisms are very complex and involve both the seed and its interaction with environmental factors.

A general theoretical model for germination strategy of seeds under varying conditions of survival has been proposed by Cohen (1966, 1968). His models take into account the germination behaviour of the seeds, their dormancy and the ability of plants derived from the germinated seeds to produce a new generation. His models do not permit a prediction of population growth.

The ecological conditions prevailing in a given habitat will affect germination. In this respect probably not overall climatic conditions but rather the micro-climatic conditions prevailing in the immediate vicinity of the seed will be the determining factors. Normally speaking, seeds are shed so as to fall either on soil or leaf litter, or in some cases they may fall in a region more or less covered with water. In other words, seeds are usually situated in or on the soil or in water. The conditions prevailing under these circumstances will depend on the nature of the soil, its chemical composition and its physico-chemical structure and on their depth in the soil or under water. Depth will influence aeration as well as penetration of light. The chemical composition of the soil, or of the water, may affect germination in a number of ways. Possibly the soil may consist of leaf litter or partially decomposed matter which contains substances inhibiting germination. The soil may have a salt content which will osmotically retard or prevent germination or it may have an ionic balance unfavourable to germination. Soil structure will, in addition to affecting aeration and water content, also determine the ability of a seedling to emerge above the soil and establish itself or, in the case of a seed falling on the soil, the ability of the roots of the seedling to establish themselves in the soil.

Although seed size and shape show a remarkable constancy, and are genetically determined, the phenomenon of somatic polymorphism is well known (Harper *et al.*, 1970). In such polymorphic seeds, produced on the same plant, different ecological roles and differences in dormancy and in germination behaviour are often associated with the different seed forms. In some cases differences in seed shape, size or weight may ensure differences in seed distribution in space. In other cases the polymorphism is probably due simply to the shedding of the seeds at different stages of development, which may result in a difference of germination in time, due to after-ripening requirements.

The softening of seeds in the burr of *Medicago* follows a definite sequence according to the position of the seed. Softening begins at the calyx end and proceeds along the spiral of the burr. This may constitute another instance of seed polymorphism, permitting distribution of germination in time (McComb and Andrews, 1974).

The adaptive value of polymorphism in the genus *Rumex* has been related to microclimatic conditions prevailing in the soil and to differences in the survival of heavy as compared to lighter seeds (Cavers and Harper, 1966; 1967). Thus the size, shape, structure and composition of seeds can determine their germination behaviour in different environments. These factors are therefore a kind of adaptation, not necessarily dependent on the germination behaviour itself. At the same time one must beware of ascribing ecological functions to all such variations, until they have been proved experimentally. We will first try to enumerate the various external factors affecting germination and their relation to various types of habitat. Following this, we will try to relate the existence of certain germination-regulating mechanisms in seeds to factors in the habitat which might be connected with such mechanisms.

## I. External Factors as They Appear in Various Habitats

### 1. *Water*

The water content of soil can vary from saturation, as in swamps and waterlogged soils, to zero or near zero for sandy soils in arid regions. Again, the precise water content of a given soil can vary with climatic conditions, coverage of the soil by plants and especially seasonally, with rainfall. The moisture content of the soil may vary within wide limits, not only between different soil types, but also in the same soil at different times of the year. Frequently habitats are classified according to their total moisture content, as hydric, mesic or xeric. However, for germination, the availability of water at a given period of time is the determining factor. Availability will be determined by osmotic factors, binding of water by soil colloids, capillary forces and soil composition and texture. In addition, of course, availability of water will be determined by competition with other organisms requiring and competing for water.

The moisture content of the soil may show seasonal periodicity. In some regions, high moisture content is associated with high temperatures, i.e. when summer rains are followed by dry winters or by very cold winters with frozen soil, and in other regions with low temperatures, i.e. when winter rains are followed by hot, dry summers. Sometimes, high moisture content occurs when the soil temperatures are at or near freezing so that, although moisture is present, it is in fact not available. Seasonal variability in moisture content is, however, by no means universal, e.g. in those regions where precipitation is more or less equally distributed throughout the year. On the other hand, where rainfall is seasonal, where summer rains or winter rains occur, the moisture content of the soil can vary from very high, immediately after heavy precipitation, to near dryness before the next rain, even within the rainy period.

A special condition exists in habitats of high salinity. Soils may be coastal, where the source of salinity is chiefly sea spray which, by causing accumulation of salt in the soil, also changes water availability. Inland saline soils are frequently marked not only by high salt content but also by being relatively impervious, having poor drainage and by being often, at any rate for part of the year, flooded. Furthermore, certain areas are characterized by periodic flooding, the plants growing in the periods between successive floodings. A very special case is constituted by the mangroves which actually grow with their root-systems submerged in sea water. On the other hand certain regions, such as some areas around the Dead Sea, have a very high salt content but are practically dry throughout the year because of absence of rain and relatively high temperatures all the year round.

## 2. *Temperature*

The soil temperature is extremely variable and shows both diurnal and seasonal changes. The extent of the changes depends on the one hand on the type of the soil (heavy or light, with or without leaf litter) and on the other hand, on the climatic conditions prevailing. In addition, very steep gradients of temperature with depth usually exist. These again will depend on the type of the soil. Soil texture and structure as well as the amount of water present in the soil, conditions for evaporation of water from the soil, and plant cover, all play a part in determining soil temperature. Usually, the upper layers of the soil show wide fluctuations and as greater depths are reached, conditions become more and more constant throughout the year.

## 3. *Gases*

The composition of the gaseous phase in the soil can be variable. Oxygen, nitrogen and carbon dioxide are the three gases normally present. As the equilibrium between the air above and the gaseous phase in the soil is attained only very slowly, appreciable differences between the phases may exist. The lighter the soil, the lower its organic matter content and the smaller the number of micro-organisms in it, the greater will be the correspondence between composition of the air in the soil to that above it. In soils having a high organic content and containing many micro-organisms the carbon dioxide content may be very much higher and the oxygen content much lower than in the air. In water-logged soils and especially in heavy soils the oxygen content of the gaseous phase may drop considerably below that normal in the atmosphere. This is also true, in general, for soils having appreciable vegetation. In such soils the roots of the plants will take up oxygen and produce carbon dioxide, again changing the balance of gases.

The volume of the gaseous phase differs greatly in different soils, being greatest in non-compact soils and least in heavy, compacted soils or water-logged soils. In addition to the three main gases, the soil may contain others, chiefly due to the activity of micro-organisms and the absence of oxygen. Thus, soils may contain methane, hydrogen sulphide, hydrogen, nitrous oxide and probably also small amounts of carbon monoxide and ammonia.

## 4. *Light*

Light is abundant usually only on the surface of the soils. In light or sandy soils light penetrates a short distance into the soil, although its intensity falls off rapidly. In heavy soils light hardly penetrates at all. In cases where the soil is covered by water, light will penetrate considerable distances provided the water is clear.

Light intensity will fall off rapidly under a vegetation cover and its spectral composition is liable to change due to differential absorption and reflection under a canopy.

## 5. *Biotic Factors*

Seeds in their natural habitat interact with other plants and with animals. The interaction with other plants may be due to inhibitors, stimulators or modification of the micro-habitat. Animals may affect germination behaviour, for example by seed softening in the digestive tract or due to distribution to other habitats. Man, using range management, or causing pollution or other technological changes in the environment, can be an important contributor to biotic factors. Fires both accidental or planned can also affect germination behavior.

In germination any two or all of the factors mentioned can show marked interaction. We will now attempt to see what part these various external factors can play in regulating germination.

## II. Ecological Role of External Factors

### 1. *Moisture and Temperature*

Probably the most crucial factor in determining germination of seeds in the soil is a suitable combination of temperature and moisture. As seeds are a means of propagating the species, their germination should occur at a time which will favour survival of the seedling. This time very often does not coincide with the time of shedding of the seeds. Despite the fact that moisture is often present when the seeds are shed, the seeds may not germinate. This may be because they have some special temperature requirement which postpones germination for some time after the seeds have been shed. The value of such a temperature requirement is obvious in the case of plants which shed their seeds in autumn. Immediate germination would lead to the killing of the seedlings in the subsequent cold winter. Postponement of germination by some sort of dormancy, which may be due to a temperature requirement or some other factor, will gi· e the developing seedling a better chance for survival. Went (1957) cites examples of this type of behaviour for a number of plants growing in the Colorado desert. When soils from this desert were moistened in the laboratory at low temperature (10°C), chiefly winter annuals germinated. If they were moistened at higher temperatures (26–30°C) only summer annuals germinated, while at intermediate temperatures yet a third group of plants could be recognized. These groups corresponded very well to those actually occurring in the natural habitat. The flora resulting from summer rains included such plants as *Amaranthus fimbriatus*, *Euphorbia micromera*, *E. setiloba* and *Portulaca oleracea*. The flora after winter rains included various species of *Gilia*, a variety of small *Oenothera* species, including *O. palmeri* and *Mimulus bigelovii*. The third group which occurred at intermediate germination temperatures in the laboratory, corresponded in nature to plants germinating after rains falling in very late autumn or early winter. Winter rains occur very regularly each year while the summer rains are far more infrequent. Apparently, therefore, the summer annuals must have a fairly extended viability as a condition for survival, a condition which is not an obvious essential for the winter annuals. Some seeds in this desert will only germinate very shortly after ripening, e.g. *Yucca brevifolia*. Apparently in this plant, if the seeds fail to germinate, they are rapidly destroyed by attacks of rodents and insect larvae. Usually only very few seem to germinate at all, and then only if summer rains follow immediately after the seeds are shed.

Similar observations were made on some of the plants of the desert region of Western Australia (Mott, 1972). Following summer rains grasses predominated in the vegetation, while following winter rains mostly dicotyledons were found. The observed germination behaviour and seedling establishment could be reproduced in controlled experiments, when top soil containing the seeds was incubated in the laboratory under different conditions of temperature. A combination of moisture and temperature conditions in the top soil determined germination behaviour. Such combinations may often be important in determining germination and survival of the species.

A requirement for a fixed temperature is not an invariable rule. Often alternating temperatures are required which constitute an adaption to certain climatic conditions. Juhren *et al.* (1953) investigated the germination of various grasses and especially of a number of species of *Poa*. They found that although all the species they studied could germinate and develop at moderate temperatures and with moderate diurnal alternations of temperatures, only one species, *Poa pratensis*, could stand diurnal changes of 26°C by day, 20°C by night in association with summer sunlight. This condition is one which the plant in question would occasionally meet in its normal habitat. On the other hand it did not germinate at all in the cold. *Poa scabrella* and *P. bulbosa* apparently were adapted to germination in the cold, day and night 3·5°C, or day 6°C and night 3·5°C. These plants occur in regions where cold conditions do prevail. *Avena fatua* seems to germinate particularly well after frequent alternations between freezing and thawing, followed by periods of wetting and drying. This treatment corresponds to the climate in those periods in the year in Bavaria when massive infestations by this weed occur (Bachthaler, 1957).

The effect of temperature cannot be readily isolated from that of moisture. However, clear-cut adaptations of germination behaviour to special climatic conditions of temperature are known. Seeds of plants from the Mojave and Sonora deserts germinate rapidly after treatment at 50°C, while lower temperatures result in long delays till full germination occurs (Capon and van Asdall, 1967), an apparent adaptation to high temperature prevailing in desert areas, prior to rainfall. The seeds of *Silene secundiflora*, when freshly harvested germinate only at 7–16°C, but after-ripening for 4 months broadens the temperature range of their germination. This plant is restricted in its range of distribution. The temperature range of its germination and changes in it appear to ensure non-germination after shedding and germination in about September (Thompson, 1971a). Good correlation between geographical distribution and temperature requirements for germination have been observed in various species of Caryophyllaceae, but transposition of a given species, by man, from one geographic area to another did not result in changes in its germination behaviour and temperature requirements. This has been shown convincingly for *Agrostemma githago* (Thompson, 1970, 1971b). However, different populations of the same species of *Mimulus* did show adaptations in their germination requirement in accordance with temperature conditions prevailing in different areas (Vickery, 1967). Thus temperature alone can be an important ecological factor in seed germination.

In a number of plants germination only occurs if the seeds are shed in a very moist habitat. An example of such seeds is willow, which is characterized by a very rapid loss of viability. The presence of the trees in very moist habitats may therefore be related to the property of the seeds. Seeds of *Saponaria officinalis* are dormant at maturity but lose their dormancy after submergence for 2 months under water, followed by exposure to air (Lubke and Cavers, 1969). This plant grows normally in gravel banks of rivers.

Seeds of many plants will not normally germinate under water and frequently, if such seeds are kept under water for any length of time, their viability becomes impaired. Other species can withstand being placed under water and will even germinate under such conditions. Frequently, germination is greatly improved by increased aeration. The moisture tension under which different plant seeds will germinate can be very different. Some species will germinate under a wide range of conditions of moisture tension, while others are much more specific in their requirements. The number of seeds which germinate under water is comparatively small. It includes such species as *Rorippa nasturtium* (water cress) and *Typha latifolia* (Morinaga, 1926). Again this requirement seems to be not under-water conditions as such, or for the presence of an excess of liquid water, but a requirement for the anaerobic conditions which are associated with this. The case of *Typha* has been considered in some detail by Sifton (1959). This plant will normally germinate under water if given a light stimulus. The seeds will germinate not only under water but also in normal conditions, e.g. on moist filter paper, provided the oxygen tension is lowered (Table 7.1). Sifton found that germination is not simply conditioned by light and

Table 7.1—Germination of *Typha latifolia* L. Seeds under Varying Conditions
(Compiled from data of Sifton, 1959)

|  | % germination | | |
|  | Seeds immersed in water | Seeds on moist blotting paper | |
|  |  | In air | In 2% $O_2$ |
| --- | --- | --- | --- |
| 35°C Light | 81 | 48 | — |
| 30°C Light | 89 | 61 | 96 |
| 30°C Dark | 4 | — | — |
| 25°C Light | 86 | 44 | — |
| 20°C Light | 69 | 37 | — |
| 15°C Light | 24 | 20 | — |

oxygen tension but also by temperature. The general conclusion reached from this study was that the germination of *Typha* was determined by the amount of water which can be held by the colloidal protein of the aleurone grain. The water-holding capacity of these grains is increased by illumination with white light and by products of anaerobic processes. However, if respiration is very vigorous, vacuolation in some of the cells is said to occur and the resulting turgor pressure can compensate for the lack of swelling of the aleurone grains, thus again permitting germination. This would account for the observation that some of the seeds will germinate even in air.

An adaptation of germination and seedling survival under anaerobic conditions also seems to exist in some varieties of rice (see Chapter 3), but in rice, at any rate, this is not an obligatory habitat or even one desirable for seedling development, while in *Typha* anaerobic conditions were the preferred ones.

Instances are known where submergence of seeds in water, followed by exposure to air, promotes germination. Seeds of both *Heliotropium supinum* and *Mollugo hirta* will germinate only if they are buried in wet mud for a certain period of time and then exposed to air. These seeds have an additional requirement for germination, viz. low temperature for a period of time (Mall, 1954). It appears that continuous soaking increases the permeability of the testa. Both these plants occur in the drying pools and puddles which form following monsoon rains. Thus it might be possible that the conditions required constitute an adaptation to a very specific habitat. However, another plant, *Polygonum plebejum*, seems to occur in the same habitat and yet does not require pre-soaking for its germination.

In wild rice, *Zizania aquatica*, a much more complex adaptation exists (Simpson, 1966). The seeds germinate in a habitat undergoing extreme temperature variations, which necessitates some degree of dormancy. The mature seeds normally drop to the bottom of a lake bed, where they remain for long periods, partly under lake ice. Dormancy is lost under low $O_2$ tensions, and at low temperatures, i.e. under those conditions which prevail in the period between the shedding of the seeds and the time favourable for their germination, in spring. Drying the seeds in air leads to a rapid loss in viability.

The sensitivity of seeds to water tension is determined in some cases by the presence or absence of a mucilage cover (Harper and Benton, 1966). Seeds of *Lepidium sativa* are not sensitive to water tension, *Plantago major* shows some sensitivity, *Reseda alba* even more, while *Vicia faba* which is devoid of mucilage shows considerable sensitivity to water tension. Mucilages may function in a different way. The mucilaginous hairs of *Blepharis persica* actually serve to orient the seed, when wetted, at an angle ensuring contact of the radicle with the soil (Gutterman *et al.*, 1967). Excess moisture prevents germination, apparently due to the creation of a diffusion barrier to oxygen by the imbibed mucilage. This species is characterized by very rapid germination in wadi beds, where moisture conditions favourable for germination and survival are probably of very brief duration. The orientation of seeds towards the soil is frequently determined by morphological features of the seed, for example in the Compositae (Sheldon, 1974).

An entirely different problem exists in plants growing in arid regions. In these plants, survival of the species is determined by mechanisms which ensure that germination occurs at a time when the seedling will be able to establish itself. Thus, ideally, germination should occur when moisture and temperature conditions favour both germination and seedling growth. A variety of mechanisms has at different times been suggested as operating in the seeds of various plants growing in such conditions. However, some advantage might result from the mere prevention of germination under exceptionally unfavourable conditions. A survey of the known germination behaviour of desert seeds does not give any very clear picture, but germination inhibitors seem to play a role. Some of the earliest experiments were those of Went. He suggested that in certain desert seeds germination only occurs if the rate of reformation of an inhibitor in the seeds, when they are moistened, is slower than the process of germination itself. Such situation might occur only under very specific conditions of moisture and temperature (Went, 1953 and 1957). Thus in seeds of *Pectis pappoza* germination occurs only after 25 mm of summer rains and not after an equal or greater amount of rain in the winter. A different regulatory mechanism connected with rain-fall has been described by Soriano (1953) for *Baeria chrysostoma*. Seeds of this plant germinate only if they are sown on wet soil. However, the subsequent development of the seedling was found to be dependent on the amount of additional precipitation which the seeds received. In the absence of additional precipitation, total germination was lower in the presence of an added 50 mm of rain immediately after sowing on wet soil.

Went, on the basis of these and many other instances of the behaviour of desert seeds, believes that the seeds contain one or more inhibitors which separately or together act as a kind of rain gauge which determines when germination will occur. It is supposed that the inhibitor content at which germination occurs is correlated to the amount of rain-fall which will permit seedling establishment. Evidence for the presence of inhibitory substances in seeds or fruits of certain desert plants is available (Koller, 1955; Koller and Negbi, 1959). However, it is unfortunate that the nature of these inhibitors has never been ascertained. Moreover, quantitative changes in them have never been proved experimentally. Thus, attractive as this theory is, it is still lacking final experimental proof. Undoubtedly this mechanism is not universal and others also occur. Thus Koller and Cohen (1959) showed that seeds of three *Convolvulus* species occurring in arid areas are impermeable to water. The seeds germinate only if their permeability is raised mechanically by abrasion or impaction, or chemically, by acid treatment. In addition, a fairly high temperature is also required. Lack of permeability prevents germination. However, the seeds become permeable gradually over a long period of time. It is, therefore, possible that each year there will be a certain number of seeds which can imbibe water and subsequently germinate, when temperature conditions become favourable. In this way not all the seeds will germinate together. If, subsequent to germination, the seedlings are destroyed, the entire population will not be wiped out, as part of the seeds will still be impermeable and viable. Survival of the species may be ensured even if germination is only restricted by lack of permeability to water and is not regulated as such. Impermeable seed coats are of very frequent occurrence in many seeds, especially among the Leguminosae. This characteristic of the seed coat does not appear to be restricted to plants occurring in dry habitats. The germination of *Calligonum comosum*, a desert shrub growing in the Sahara and similar habitats in nature appears to be limited to sandy environment. The germination behaviour in this case seems to restrict the plant to such habitats as coarse sandy soils with low rainfall. This appears to be due to the fact that the seeds will not germinate if in contact with liquid water for any length of time. Again the germination behaviour is complicated by light sensitivity of the seeds, which germinate better in the dark than in the light. Under natural conditions germination seems to occur only if the seeds are buried in sand (Koller, 1956).

Another type of adaptation to moisture is met within the seeds of *Panicum antidotale*, *P. turgidum* and *Atriplex dimorphostegia* (Koller, 1954). Seeds of these plants germinate to a higher percentage in the laboratory if the seeds, together with their dispersal unit, are dried over calcium chloride before sowing. In their natural habitat the seeds are often exposed to hot dry weather particularly in the direct sun, before they germinate. The mechanism of pre-drying might thus constitute a special adaptation to their normal habitat. It is, however, not clear what is the precise relative humidity in the microclimate surrounding the seeds. Pre-drying is certainly not an essential for germination, as some of the seeds germinate even without it and the whole mechanism only points to a correlation between habitat and germination behaviour.

## 2. *Gases*

From an ecological point of view, the function of gases in regulating germination is very puzzling. The normal requirement is for aerobic conditions, i.e. 20 per cent oxygen in the atmosphere. As discussed previously, germination of many seeds is prevented when the oxygen content is lowered appreciably. However, the seeds of certain plants and particularly of aquatics such as *Typha* prefer anaerobic conditions for germination. Morinaga (1926), when testing a variety of seeds, found that among 70 species at least 43 were capable of germinating under water, i.e. under anaerobic conditions; 18 out of these 43 species germinated equally in water and in air on moist filter paper. Only two species had a definite preference for anaerobic conditions. The only natural habitats where one would expect to meet anaerobic conditions are under water, and in waterlogged soils and swamps. Nevertheless, in all kinds of habitats the composition of the atmosphere surrounding seeds can change. The ratio of oxygen to carbon dioxide in particular is liable to change, for example, as a result of the activities of the microflora and fauna in the soil. This may lead to an increase in the carbon dioxide, concentration with a simultaneous decrease in the oxygen content of the atmosphere in the soil. A different way in which the gaseous environment of the seeds may change arises from the differential permeability of the seed membranes to different gases. This can lead to an increase of carbon dioxide and a decrease of oxygen within the seed. Both in the soil and in the seed the changes in the gaseous atmosphere are confined to the oxygen and carbon dioxide fractions, as the nitrogen content usually remains constant.

No precise information of the internal atmosphere of the seed is available. The only attempt to measure this seems to have been made by Kidd (1914) in peas (Table 7.2).

Table 7.2—The $CO_2$ Content of Peas, *Pisum sativum*, While Germinating (After Kidd, 1914)

| Time of germination (hr) | $CO_2$ content ml $CO_2$/100 g of seeds |
|---|---|
| Dry Seeds | 145 |
| 18 | 64 |
| 25 | 41 |
| 39 | 43 |
| 64 | 39 |
| 97 | 16 |

Kidd ground peas either in water or in barium hydroxide. Those ground with water were allowed to equilibrate with the air and then barium hydroxide was added. The difference in titer between the two treatments was taken as a measure of the carbon dioxide content of the seeds. This method must be regarded with considerable reserve as regards the carbon dioxide content of the seeds, and does not indicate the carbon dioxide/oxygen ratio within the seed. All other conclusions on the gaseous atmosphere in the seeds are based on the response of seeds, with or without seed coat, to changes in the composition in the external atmosphere. This need not always correspond to a change in the internal atmosphere.

Some information is available about the permeability of isolated membranes from seeds, as described earlier (Chapter 4). Kidd concluded from his experiments that internal carbon dioxide accumulation, above a certain level, inhibits germination. Germination will occur only when this level falls. According to Kidd such a drop occurs in peas about 18 hours after the seeds are placed in water (Table 7.2). If, in fact, this conclusion is correct, then this may possibly regulate germination. As long as internal carbon dioxide concentrations are high, germination is retarded, even if the seed is imbibed. This may even induce secondary dormancy. When the impermeable membrane is destroyed or punctured, the accumulated gas is released and germination follows. This could cause a spread of germination over a period of time, depending on the occurrence of damage to the seed coat.

However, Thornton (1944) showed that a large number of seeds, particularly of crop plants, germinate in carbon dioxide concentrations as high as 40–80 per cent, provided 20 per cent oxygen is also present. He concluded that for the seeds used, carbon dioxide does not inhibit germination in the presence of oxygen. In fact, Kidd also found that increasing oxygen concentrations and increasing temperatures decreased the inhibition caused by carbon dioxide, which is confirmed by Thornton. Under natural conditions a situation is rarely, if ever, met with, where carbon dioxide concentrations rise and yet the oxygen content remains high. Thus the conclusions reached by Kidd for *Brassica alba* seem to correspond more closely to events occurring naturally than the examples quoted by Thornton. In *Brassica* seeds the carbon dioxide was assumed to decrease the permeability of the testa, especially towards carbon dioxide. For *Cucurbita* (see Chapter 4) it has been shown that the seed membrane which controls gaseous diffusion is more permeable to carbon dioxide than to oxygen. Thus in *Cucurbita* it is possible that this permeability effect ensures spread of germination over a period of time.

In a number of seeds, small amounts of carbon dioxide can promote germination. The dormancy-breaking action on *Medicago* and *Trifolium* was mentioned in Chapter 4. In these plants this response may have resulted from a process of selection by agricultural practice rather than from natural selection. It seems possible that the sensitivity to small amounts of carbon dioxide causes rapid germination and that plants of this type are selected for, as this is often a desirable property in agricultural crops.

A suggested difference between weed seeds and the seeds of crop plants is their ability to survive while buried in soils for long periods. Under these conditions the seeds are liable to be exposed to high carbon dioxide concentrations while imbibed. Weed seeds are not damaged by these conditions and germinate rapidly when removed from the soils, as for example after ploughing. The seeds of many crops seem to suffer under these conditions. However, some weed seeds lose their viability quite rapidly if buried in the soil. For example *Scandix pecten, Linaria minor, Bartsia odontites* and *Polygonum aviculare* showed a reduction of 90 per cent in their viability in a 2-year period (Brenchley and Warington, 1933). In the seed burial experiments of Duvel (1905) it was found that *Avena fatua* and *Lactuca scariola* survived under conditions where *A. sativa* and *L. sativa* lost their viability. Similar differences were observed between seeds from wild and cultivated plants of *Helianthus annuus*.

Relatively little direct information is available about the effect of oxygen on germination. Proof of impermeability or partial permeability to oxygen comes from the work of Marchaim *et al.* (1972) and of Come (1970). Indirect evidence for impermeability to oxygen exists for the testa of the upper seed of *Xanthium*. A rise in the external oxygen tension raises germination. In these seeds, therefore, the testa, if not impermeable to oxygen, is at any rate only partially permeable to it. However, as already discussed, under natural conditions oxygen concentrations do not rise above 20 per cent. The ecological significance of these findings is not entirely clear. The germination of *Xanthium* seeds seems to occur only when a certain minimal threshold value of the internal oxygen concentration is reached. This results in the oxidation of an inhibitor present in the seeds. This inhibitor apparently prevents germination. Nevertheless, although there is a good correlation between external oxygen concentration and germination, it is not absolutely certain that germination is the direct result of destructions of the inhibitor (Wareing and Foda, 1957; Porter and Wareing 1974). Under natural conditions the necessary threshold value is attained only very slowly due to the low permeability of the testa to oxygen. If the seed coat is punctured or damaged in some other way, these processes will be greatly accelerated. If the seed coat is completely impermeable to oxygen, then only damage can enable germination (Crocker, 1906; Shull, 1911).

The work of Edwards (1969) discussed in Chapter 4, indicates a role of oxygen in controlling the concentration of a germination inhibitor in seeds of charlock (*Sinapis arvensis*). This could be a more general phenomenon by which the external oxygen concentration via its effect on the internal oxygen concentration controls germination. Come (1970) has also indicated that in apple seeds a low oxygen concentration induces secondary dormancy, possibly because of the interaction of oxygen with germination inhibiting substances in the seed. Although this kind of result provides a possible explanation of the mechanism by which oxygen controls germination it does not account for the ecological role of oxygen. In general it may, however, be stated that low oxygen concentrations are usually accompanied by either an excess of water, or a lack of light, or both. In all probability the ecological role of oxygen must be sought under those conditions under which one of these two additional factors is of importance.

From the above it is clear that no definite conclusion can be drawn about the function of gases in regulating germination. It seems probable that the internal concentration of gases in the seed is the only determining factor and it is possible that the carbon dioxide/oxygen ratio is determining rather than the absolute concentrations of each.

## 3. *Light*

An obvious advantage arising from the regulation of germination by light would be, in seeds having different light requirements, in adapting them to their habitat. Thus, germination of seeds requiring light may be prevented when they are buried under soil or leaf litter, or promoted when they fall on the soil surface. Such behaviour may determine how well a seedling will subsequently be able to establish itself. As already discussed, seeds can be divided into groups, viz. (1) those which require light, (2) those which are inhibited by light (Table 7.3 and see also Table 3.8) and (3) those which are

Table 7.3—Classification of Some Seeds According to Their
Light Requirement
A—Seeds whose germination is stimulated by light
B—Seeds whose germination is retarded by light

| A | B |
|---|---|
| *Arceuthobium oxicedri* | *Bromus sp.* |
| *Daucus carota* | *Datura stramonium* |
| *Elatine alsinastrum* | *Lycopersicum esculentum* |
| *Ficus aurea* | Liliaceae, various species |
| *Ficus elastica* | *Nigella sp.* |
| *Gloxinia hybrida* | *Phacelia sp.* |
| Gramineae, various species | *Primula spectabilis* |
| Gesnericeae, various species | |
| *Lactuca sativa* | |
| *Lobelia cardinalis* | |
| *Lobelia inflata* | |
| *Loranthus europaeus* | |
| *Lythrum ringens* | |
| *Mimulus ringens* | |
| *Nicotiana tabacum* | |
| *Nicotiana affinis* | |
| *Oenothera biennis* | |
| *Primula obconica* | |
| *Phoradendron flavescens* | |
| *Raymondia pirenaica* | |
| *Rumex crispus* | |
| *Verbascum thapsus* | |

indifferent to light. Ecologically speaking, it would be expected that all seeds requiring light for their germination belong to species of plants which necessarily germinate on the soil surface. All those seeds which are inhibited by light should belong to species of plants whose seeds will only germinate if they are covered by a certain amount of soil. However, this would be a gross over-simplification. Frequently short illuminations stimulate germination while prolonged illumination inhibits it, e.g. in *Amaranthus blitoides* and *Atriplex dimorphostegia*. Furthermore, seeds of the same genus may be light-requiring or light-inhibited, e.g. *Primula obconica*, which is light-stimulated, and *Primula spectabilis* which is light-inhibited. In more extreme cases different varieties of the same species may be light-requiring or light-indifferent, for example *Lactuca sativa*. The possible value of a requirement for short illumination may be as follows. Seeds which are entirely uncovered will fail to germinate as continuous light inhibits them. However, if the seeds are partially covered, they may be exposed briefly to light at certain periods of the day. Under these conditions the seeds will germinate. In many seeds light sensitivity is confined to only one part, e.g. in *Phacelia* sensitivity is localized at the chalazal and micropylar ends, in lettuce only the micropylar end of the seeds is light-sensitive. Such sensitivity could perhaps ensure germination of the seed only in certain orientations in the soil. It is possible that these requirements are in some way related to the micro-climate which is associated with certain conditions of illumination. The latter will affect subsequent development of the seedling. This is borne out by the complex interactions of light with other environmental factors which have already been discussed. There is no evidence, however, to show that localization of sensitivity has ecological importance. Light sensitivity is often induced in seeds. The seeds of *Spergula arvensis* and *Stellaria media* do not require light for germination when they are fresh. After burial in the soil a light requirement is induced (Wesson and Wareing, 1969). The ecological role of this requirement is presumed to be a prevention of germination till the seeds again reach the soil surface. Light induction has also been reported for *Capsella bursa pastoris* and *Senecio vulgaris*.

Light and temperature frequently interact. This interaction may be such as to make seeds sensitive to light only at certain temperatures but not at others. Some of the possible combinations of light and temperatures are shown in Table 7.4. These

Table 7.4—Classification of Light–Temperature Interactions of Some Seeds
(Compiled from data quoted by Stiles, 1950 and Koller, 1955)

| Plant | Temperature | Light | | | Dark | | |
|---|---|---|---|---|---|---|---|
| | | Low | High | Alternating | Low | High | Alternating |
| *Veronica longifolia* | | + | − | − | − | + | + |
| *Epilobium hirsutum* | | + | − | − | − | + | + |
| *Poa pratensis* | | + | − | − | − | − | + |
| *Rumex crispus* | | + | − | − | − | − | + |
| *Apium graveolens* | | + | − | − | − | − | + |
| *Oryzopsis miliacea* | | + | − | − | − | − | + |
| Gesnericeae species | | + | − | + | − | − | − |
| *Ranunculus sceleratus* | | − | − | + | − | − | − |
| *Chloris ciliata* | | × | + | − | − | − | − |
| *Amaranthus retroflexus* | | × | + | − | − | − | − |
| *A. lividus* | | × | − | − | − | − | − |
| *A. candatus* | | × | + | − | − | − | − |

+ signifies germination is promoted by combination of treatments.
− signifies no promotion.
× germination inhibited by this set of conditions.

interactions are so complex that it is almost impossible at present to interpret their ecological significance, if any. It must be borne in mind that very little is known about the micro-climate prevailing in the immediate vicinity of the seeds. In addition, the apparent combination of a number of factors, revealed in the laboratory, may not in fact exist in nature. They may be no more than residual genetic properties which no longer have any direct survival value and which are retained as long as they have no harmful effect. Stoutjesdyk (1972) suggests that such features may be part of a pool of unutilized properties on which natural selection may act in the future. It seems possible that any ecological or survival value which light may have in regulating germination exists only in very special cases. Light requirement is frequently associated with small seeds, which are supposed to contain rather small amounts of reserve materials. It is frequently assumed that, because the seeds are small and therefore they contain little reserve materials, they must germinate under conditions where photosynthesis occurs very soon after germination. But seed size is no measure of the amount of storage materials present relative to the requirements of the seedling. What determines the adequacy of the reserve materials is the ratio between the amount of these substances and the size of the embryo or seedling to be nourished by them. Moreover, a light requirement is by no means a universal phenomenon associated with all small seeds, although no large seeds are known to be light-requiring. The view outlined seems, therefore, to be an oversimplification of the ecological function of light.

Nevertheless, at least in some cases, the light requirement may be related to seedling establishment. Seeds of *Narthecium ossifragum* are light sensitive, but the light requirement is such that 10 per cent of the intensity of day light induces germination. However, subsequent growth of the seedling requires much more light. The germination behaviour of this species is complicated by the interaction of light and moisture requirements. Germination requires a high water table. However, excess water causes both a loss of viability of the seeds and damping off of the seedlings (Summerfield, 1973). Such conditions constitute a rare combination in nature. The main mechanism in this species to ensure survival is the production of a very large number of seeds to counterbalance the high rate of failure in seedling establishment.

A special case of adaptation exists in aquatic plants. In these, germination in deep water would be unlikely to lead to seedling survival. If the seeds are placed in a depth of water into which light can still penetrate, they will germinate and this will also occur if the water level falls, for example on river banks where the level varies seasonally or in swamps. This type of mechanism could explain the light requirement of seeds of *Typha latifolia*. Other plants in which this type of mechanism may be functioning are *Scirpus, Eichhornia crassipes* and *Juncus maritimus*.

Nevertheless, in general the light requirement of many seeds is difficult to understand. Particularly as in many cases, the amount of light required to stimulate germination is so small that it may be supposed that the seeds receive it under almost all normal circumstances.

It is generally assumed that the light response of seeds is mediated by phytochrome. Some recent findings try to relate the R–FR mechanism in regulating germination to its possible ecological role. The distribution of red and far-red light of incident radiation changes during the day. The incident radiation becomes enriched with far-red as compared to red at dusk and at dawn. Moreover, the spectral composition of radiation changes as it penetrates through a leaf canopy. Red light is selectively absorbed, and under a corn canopy, for example, there is a several fold increase in the relative far-red irradiance compared to red. The change in the R–FR ratio is greatly increased at lower angles of elevation of the sun, i.e. at dawn and dusk (Sinclair and Lemon, 1973). This then would indicate that the R–FR ratio does vary under natural conditions. Direct attempts to investigate the significance of such changes were made by Stoutjesdyk (1972). His experiments show that in a number of species germination was inhibited when the seeds were sown under a canopy of *Crataegus monogyna*. This applies both to seeds whose germination is stimulated by light such as *Betula pubescens* and *Epilobium hirsutum* and those whose germination is inhibited by light such as *Bromus tectorum*. He sowed seeds either under the *Crataegus* canopy or just outside it, but shielded from direct sunlight and also ran suitable controls in the dark. He also examined the spectral composition of the light reaching the seeds. The light under the canopy was enriched in FR while that outside the canopy contains less FR and relatively more blue than direct sunlight. In the same genus different species responded differently, e.g. the genus *Sagina*. The results suggest that it is possible that the ratio R–FR of light under natural conditions could affect germination and that when the FR content of the radiation is high germination inhibition may result. Presumably inhibition of germination will depend not only on the R–FR ratio of the radiation but also on the amount of phytochrome in the seeds and their sensitivity to the $P_R/P_{FR}$ ratio in them. Thus when considering the ecological role of light it is no longer sufficient to consider whether seeds are exposed to light, but the spectral composition of the light reaching them as well as the phytochrome reaction in the seeds must be taken into account.

Lastly the results indicating that the spectral composition of the light falling on the parent plant can effect the subsequent germination of the seeds must be mentioned (Shropshire, 1973).

Thus the data presently available on the response of seeds to light in their natural environment are quite inadequate in order to make clear ecological interpretations possible. The factors regulating $P_R/P_{FR}$ ratios in peas seem to be much more complicated than was thought at one time. They are determined not only by light but by the redox state of the tissue (Kendrick and Spruit, 1973), which is at least in part regulated by the oxygen tension. The ecological function of the R–FR remains puzzling, but perhaps will become clearer as more detailed work is carried out.

## 4. Soil Conditions

The ionic composition of the soil also frequently affects germination. Some seeds respond to the calcium content of the soil in their germination behaviour, while others respond particularly to sodium content.

Seeds of *Hypericum perforatum* do not germinate in soils containing more than traces of calcium, whereas tomato seeds seem to be indifferent to the calcium content of the soil.

*Avena fatua* germinates well in sandy loam and loess clay which also favour the growth of the seedling. In peaty soils germination is good but growth poor. Calcareous soils were found to be least suitable for germination among the soils examined (Bachthaler, 1957).

The high salt content of soils, especially of sodium chloride, can inhibit germination, primarily due to osmotic effects. In such saline environments the development of the seedling is extremely poor. However, a number of plants, which have a special resistance to salt, can develop. This type of vegetation is frequently termed halophytic. Many of the plants so classified are distinguished by a high salt tolerance in various stages of their development, including germination, rather than by a positive salt requirement; high salt tolerance is not therefore proof that the plant is halophytic. But some plants show a definite requirement for a certain salinity and these germinate better in the presence of low concentrations of salt, and their subsequent development is also better. This is true for *Atriplex halimus* for example (Fig. 7.1). Even in this plant

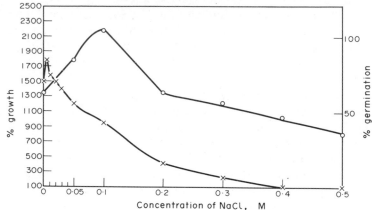

Fig. 7.1.  Effect of NaCl concentration on growth and germination of *Atriplex halimus*.
(Poljakoff-Mayber, unpublished)
×——× % germination
○——○ growth, final dry weight as % of initial dry weight

the tolerance to sodium chloride during growth and development of the seedling is ten to a hundred times greater than during germination. However, in *Atriplex* salinity improved growth only under conditions of high evaporative demand of the atmosphere; under humid conditions any addition of salt retarded growth (Gale *et al.*, 1970). It is possible that the response of germination to salinity is also dependent on some additional factor, but this has not been proved experimentally.

In many plants of this type, germination occurs when the salt content of the habitat has reached its lowest level, e.g. towards the end of or after the rainy period. However, some plants can germinate under conditions of very high salinity. Seeds of *Arthrocnemum halocnemoides* Var. Pergranulatum are still able to germinate in solutions containing 20 g/l NaCl, although the germination percentage was reduced to a half of that in water (Malcolm, 1974). The seeds for these experiments were taken from plants growing in saline water-logged areas of Western Australia. The water table is 1–1·5 m deep, and the ground water contains between 14 and 55 g/l soluble salts. It appears that these seeds are adapted to germinate and grow under highly saline conditions.

In a few plants occurring in a saline habitat vivipary is observed. The best-known case is probably that of the mangroves, which grow for most of the year directly in sea water. It seems that during the process of germination these plants are much more sensitive to salt and are more easily damaged by it than the already grown seedling. In this case vivipary may be regarded as a means of evading the unfavourable environment during germination.

Salt frequently accumulates in the organs of plants growing in saline habitats. As a result the fruits of the dispersal units of such plants often contain salt. In these plants germination is to some extent controlled by the salt content of the fruit, and the salt serves as an inhibitor of germination. In other seeds more specific inhibitors, not acting through osmotic effects, are thought to be present.

It should be remembered that soils are quite heterogeneous and germination and seedling establishment may be determined by micro-heterogeneity in the soil.

In germination studies there is a discrepancy between the germination percentage achieved under laboratory conditions and that achieved, for the same stock of seeds, in the field.

Apparently under field conditions the moisture in the surface layer of the soil is continuously depleted due to evaporation. As a result the germinating seed has to compete with the atmospheric evaporative demand for water. This competition is severe under dry-farming conditions and probably even more severe under natural conditions.

A number of different definitions have been suggested to describe the optimal conditions for soil-water relationships in the micro-environment of a germinating seed. One definition assumes that there is a critical soil water potential permitting germination, which is characteristic of each species (Hunter and Erickson, 1952). Another definition is based on the seed-soil contact. It was shown that when seed-soil contact was poor, the germination percentage was reduced even at high water potentials of the soil (Collis-George and Hector, 1966). Apparently the soil-water matric potential is the important factor. An attempt to analyse the actual conditions adjacent to the germinating seed was made by Hadas (1970) and Hadas and Stibbe (1973). They propose a model, describing the water relations of a single seed and its immediate environment as a dynamic process. Their model considers the water uptake of the germinating seed as a function of the changing factors of the soil, such as water content, water potential, water diffusivity and water flow towards the seed. From their work it appears that the germinating seed in the soil is exposed to a much more severe water stress than is measured by conventional methods in the bulk of the soil.

## 5. *Inhibitors*

Many seeds do not germinate within their fruit, as discussed earlier. This observation, made very long ago, led to the idea that inhibiting substances are present in such fruits. In the course of time this idea was further extended, when it was found that extracts of such fruits also prevent germination. This finding supported the view that failure to germinate in the fruit was due to germination inhibitors and not due to unfavourable conditions, such as lack of oxygen or light, within the fruit. The prevention of germination of seeds in the fruit ensures that germination will only occur if the seeds are in some way dispersed. This will occur if the fruit decomposes, is broken or damaged or if it is eaten by animals and the seeds subsequently excreted. In all these cases the dispersal of the seeds is increased and therefore the chances that the seeds will reach areas removed from the parent plant, where they are likely to survive, are improved. This view of the ecological function of fruits has been disputed (Gindel,

1960). Gindel claims that the presence of inhibitory substances has only been shown in fruits at a period of their development, when the seeds are not yet ready to germinate and the fruit is not completely ripe (in the biological rather than the culinary sense). According to this view, when the fruit is fully ripe, at the end of its development, many seeds will germinate in it. Examples are brought of cultivated plants such as apples, tomatoes and *Cucurbita maxima* and of a number of other plants such as *Ceratonia siliqua* (carob bean), *Laurus nobilis* and *Arbutus andrachne.* Accordingly it is supposed that during the development of the fruit the inhibitory substances present in it are converted to substances favouring germination, and that the fruit provides a nutrient medium and to some extent protection for the developing seedling. At present no evidence is available to indicate whether in fact inhibitors are decomposed during development of the fruit. Moreover, it is not clear what is the fate of the seedling after it has germinated within the fruit. Thus, no evidence exists whether the seedling remains in the fruit, without further development, until the latter decomposes or whether it pierces the fruit, as is the case, apparently, for carob beans. The case of mango may be relevant here. Among varieties of mango, viviparous and non-viviparous kinds are known. Singh and Lai (1937) found that the viviparous varieties invariably showed splitting of the endocarp at the broader end. This split endocarp was found to favour germination and thus might account for the phenomenon of vivipary. In normal fruit with intact endocarp, radicle extrusion may be prevented. How many seedlings survive this type of germination is also not clear. At present this view must therefore be regarded as highly speculative and in many instances in conflict with casual observation. The more usually accepted view had led to a search for inhibitory substances in fruits. Although these have in many cases not been identified, nevertheless correlation between inhibitory power of extracts and the germination behaviour of the seeds has been found. In fleshy fruits inhibition may be correlated with sugar content. In other instances salt content in the dispersal unit has been found to be the cause of germination inhibition, as in *Zygophyllum dumosum.* In the case of *Zygophyllum* the dispersal unit was found to contain salt in sufficient amounts to cause inhibition, but in addition weak inhibitory substances were also present which accentuated this inhibitory action (Lerner *et al.,* 1959). In a few instances specific inhibitory substances have been found to cause inhibition of germination.

Among the inhibitors identified in fleshy fruits are parasorbic acid in *Sorbus aucuparia* (Kuhn *et al.,* 1943), and ferulic acid in tomatoes (Akkerman and Veldstra, 1947) though this is probably not the only inhibitor present, while in lemon, strawberry and apricot, Varga (1957) showed the presence of a mixture of organic acids which increased in amount as the fruit ripened. In dry fruits the only inhibitor identified with any certainty is coumarin, in fruits of *Trigonella arabica* (Lerner *et al.,* 1959). Even in *Trigonella* probably other, additional, weak inhibitory substances are also present. Phenolics and coumarin and its derivatives are often reported as almost universally present inhibitors, which also can act as germination inhibitors in seed husks, coats, fruits, etc. (Van Sumere *et al.,* 1972). In the case of *Fraxinus excelsior* the presence of an inhibitor in the embryo itself has been shown by Villiers and Wareing (1960). In most of these cases the role of the inhibitors may be supposed to be that already mentioned, of dispersing seed germination over a period of time. However, the mechanism may be even more complicated. In seeds requiring chilling, the inhibitor may not cause a spread of germination in time. For example, in *Fraxinus* and *Betula* the inhibitor probably only defers germination till a suitable time of the year, when chilling creates the necessary conditions which enable germination to proceed (Wareing, personal communication).

It must, however, be remembered that these presumed functions of inhibitors in fruits are by no means finally proven, and in fact they are very difficult to prove unequivocally. It is possible to interpret the observed facts differently (as already discussed). It is to be hoped that different approaches will lead to new lines of research into the probable biological and ecological function of inhibitors. More detailed examinations of extracts from fruits and seeds have already shown that these contain a mixture of substances, some of which inhibit while others stimulate germination, while yet others are active in affecting growth. The amount of these substances changes with time and with treatment of the seeds. It seems likely that germination is not simply controlled by inhibitors but that the interaction of both the promoting and inhibitory substances regulate it, as indicated by the case of *Fraxinus*, for instance (Villiers and Wareing, 1960; see also Chapter 6).

## 6. *Biotic Factors*

Many plant organs, other than fruits and seeds, contain inhibitors of germination. It has frequently been observed that leaves or leaf litter contain compounds which can inhibit germination of a number of seeds. The accumulation of leaf litter under trees could have a regulating effect on germination. If species differences exist in sensitivity to inhibitors, then those plants whose seeds are most sensitive to the inhibitor will not germinate in the immediate vicinity of plants whose leaves or straw contain such inhibitors. The presence of inhibitors in leaf litter might affect the distribution of certain plants, favouring some species and preventing the occurrence of others. It is often observed that some plants are entirely absent in the immediate vicinity of certain other species. For example, in Israel the soil surrounding *Eucalyptus* trees is very poorly vegetated. Yardeni and Evenari (1952) showed that leaves of *Eucalyptus* contained germination inhibitors. It was suggested that the inhibitors present in the soil or leaf litter might be responsible for the relative bareness of soil near *Eucalyptus*. However, later experiments showed that although a very strong inhibitor could be isolated from the leaves, the soil from *Eucalyptus* groves did not inhibit germination of either wheat or lettuce (Lerner and Evenari, 1961). Thus the inhibitor in the leaves is either washed out or destroyed and no longer exercises a biological function. However, it is possible that very small amounts of inhibitor, together with the mechanical effects caused by the leaf litter, may be responsible for prevention of germination. For example, Dinoor (1959) showed that the leaf litter under Valonia oak constituted a mechanical obstacle to seedling establishment. When germination occurred on the leaf litter, the roots failed to reach the soil and if it occurred underneath the litter, the shoot failed to emerge. Similar effects may occur in *Eucalyptus* groves or in other habitats covered with leaf or straw litter.

Inhibitors are present in the litter of a number of plants. Thus in rice straw Koves and Varga (1958) showed the presence of a number of phenolic compounds which have inhibitory properties and suggest that these compounds may have some biological function. Beech litter also contains inhibitors. In this case inhibitors are absent in the litter immediately after the leaves are shed, but appear after it has been exposed to one winter (Winter and Bublitz, 1953). This could be related to the relatively brief viability of beech seeds, which germinate shortly after shedding and the seedlings establish themselves before the inhibitor develops in the litter.

The caution with which results of experiments with extracts of leaf litter should be regarded is demonstrated by the case of leaf litter of *Backhousia* (Cannon *et al.*, 1962). *Araucaria* fails to regenerate in regions where there is a large amount of leaf litter from *Backhousia*. Extracts of the leaf litter do in fact inhibit germination of *Araucaria*, the active compound being dehydroangustione. However, field experiments showed that on the leaf litter itself *Araucaria* germinates well. Watering of the litter increased germination. Thus, in this case the litter was acting in a different way, perhaps by causing seedling death, rather than by inhibiting germination.

Other causes for the absence of plants in the vicinity of certain trees may be the competition for water and light, which may have a more decisive effect than inhibitors. If competition is very strong, then small amounts of inhibitors may also play some part in regulating plant distribution.

Competition apparently is the crucial factor in establishment of *Juncus* species among the sward of grasses. The *Juncus* can germinate and establish itself well when the soil is very moist. However, if the soil dries out a little, then *Juncus* no longer germinates readily. Under these circumstances other plants germinate and establish a cover which prevents light reaching the *Juncus* seeds, which require light for their germination. The *Juncus* is subsequently effectively prevented from germinating (Lazenby, 1956). Such *Juncus* seeds retain their light sensitivity, even if buried, for many years.

Many members of the Cruciferae, such as *Brassica* and *Sinapis*, contain complexes of mustard oils in their fruit as well as in other parts of plants. In *Sinapis arvensis* the fruits contain mustard oils. Those seeds which are shed from the fruit germinate readily. The upper part of the fruit does not open readily and retains its single seed. This seed only germinates later, when the mustard oils have been washed out of the fruit. An interesting example of inhibitory action by other parts of the *Brassica* is provided by the case of *Brassica nigra*. Certain areas in California in the United States, which were damaged by fire, were resown with *Brassica nigra*. In the areas so treated many species failed to re-establish, while in other areas, not resown with *Brassica*, they germinated well and re-established themselves (Went *et al.*, 1952). Apparently, *Brassica* leaf litter contains an inhibitor which prevents the germination of certain plants. The presence of a water-soluble inhibitor in leaves of *Brassica nigra* was in fact proved (Scherzer, 1954).

In the aerial parts of *Echium plantagineum* two distinct inhibitors were demonstrated which are of special interest because they affected different species of plants quite differently. Differential effects were clearly demonstrated in this instance (Ballard and Grant Lipp, 1959).

The leaves of *Encelia farinosa* contain a powerful inhibitor of both germination and growth. A mulch of leaves of *Encelia* can inhibit the germination of a number of plants (Bonner, 1950). However, the suggestion that this accounts for the absence of plants very near to *Encelia* shrubs has been disputed by Muller (1953). The latter claims that *Franseria* plants growing in a similar habitat also contain powerful inhibitors, yet its immediate environment is populated by other plants. Muller ascribes the difference between the ground cover in the immediate vicinity of these two shrubs as being due to differences in the growth habit and in the formation of leaf litter by them. Under *Encelia*, shrub-dependent herbs fail to develop because little leaf litter accumulates and the plant has a short life span. Under *Franseria dumosa*, however, leaf litter accumulates and therefore shrub-dependent herbs develop. Even under *Encelia*, open ground species develop readily.

A further instance of a plant containing a germination inhibitor is *Artemisia absinthium*, the inhibitor apparently being the alkaloid absinthin. A number of plants sown in the vicinity of *Artemisia* fail to germinate or even to develop. For example, *Levisticum officinale* is killed up to a distance of one metre. Some plants are more resistant than others, *Senecio*, *Lathyrus clymena* and *Linum austriacum* being extremely sensitive, while *Stellaria* and *Datura* are quite resistant (Funke, 1943). The ecological function is again in doubt, as seedlings even of sensitive plants, which survive for a year near *Artemisia absinthium*, subsequently develop quite normally, without further inhibition. The decomposing leaf litter of *Celtis laevigata* did significantly inhibit the germination of a number of species (Lodhi and Rice, 1971). An interesting mechanism has been suggested for the seeds of *Coumarouna odorata* (Valio, 1973). Seeds of this plant have a high coumarin content and their germination is quite insensitive to exogenous coumarin. After germination the radicle of this plant excretes coumarin and this can then apparently inhibit the germination of other seeds in the vicinity. It remains to be proved that this effect occurs under natural conditions in the soil and not only in the laboratory.

Many examples are found in botanical literature of the effect of plants on each other. Many of these are based on visual observation and most experiments have not been sufficiently rigorous to permit interpretation. Although it appears that plants do excrete substances or that substances leach out from them, which can affect other plants or seeds in their environment, the magnitude of the effects and their biological importance is still not clear. This whole problem has been reviewed by Evenari (1965).

The seeds of certain parasitic plants present a special ecological problem. Both *Striga* and *Orobanche* can develop only if the seeds germinate very near to plants which they can parasitize. The germination of these parasitic plants normally requires the presence of a stimulator, although they can be induced to germinate artificially. It appears likely that the stimulator which is required by these seeds is excreted or leached out from the roots of very many plants, and that it constitutes a normal metabolite in such plants. The requirement of *Striga* and *Orobanche* for such a factor in their germination would ensure that they germinate only under conditions where the chance to parasitize a suitable host is good. However, the apparent non-specificity of the stimulator, as compared to the specificity in the host requirement of these plants, throws some doubt on such an interpretation. Moreover, it appears that there is no one single stimulatory compound. For example, Sunderland (1960) was able to show in maize roots at least one water-soluble and a number of ether-soluble substances stimulating the germination of *Striga* and *Orobanche*. These substances act to some extent synergistically. It is possible that in different plants specificity effects are due to differences in the ratio of the amounts of such compounds acting synergistically. It remains to be seen whether this could cause certain species of parasites to germinate only near their specific hosts.

In the case of *Viscum album* and other *Viscum* species nothing is known about special germination requirements, the sticky nature of the fruit merely increasing the chances of it being carried from tree to tree by birds. This increases the likelihood of the seeds reaching a suitable host. The subject of germination of parasitic angiosperm seeds has been reviewed by Brown (1965), and more recently by Edwards (1972).

Fire is another important factor which can control germination. This may be in one of a number of ways. Fire can remove vegetation and so improve light and aeration and at the same time remove competition for space, light and nutrients between the seedlings which are establishing themselves and the existing plants. Went *et al.* (1952) suggested that in addition fire can destroy accumulated inhibitors present in the soil cover and uppermost layers of the soil, again removing a possible cause of failure of seeds to germinate, or of the seedlings to establish themselves.

Many animals can change the balance of different plants in a given area by grazing, by distributing the seeds, by the excretion of seeds after eating fruits, and by other means. Goats grazing in the Middle East denuded the area of forests and as a result caused the establishment of a different type of vegetation, containing many annuals and shrubs. Insects of various kinds which collect seeds may have certain local effects, for example, ants accumulating certain kinds of seeds in or near their heaps.

Ants may play an important role in the destruction of seeds, particularly those seeds which are used for forage. They may collect seeds near their nests, some of which still may be able to germinate (Carroll and Janzen, 1973). In addition some species of seeds are specifically adapted to distribution by ants. Thus seeds of the species *Vancouveria* have a special appendage containing oil droplets, the elaiosome, which makes the seeds attractive to ants (Berg, 1972). The oil containing appendage is utilized by the ants, but the seed itself is not eaten. In this case the ants act as a dispersal agent, and apparently numerous other cases are known although few have been studied in detail.

Man can exercise and has exercised a profound effect on the distribution of plants in certain areas. Agricultural practice has always been designed to cause the establishment of certain plants to the entire exclusion of others. The disastrous results which this policy may have, was shown in the centre of the United States which was converted into an almost barren area for many years, the so-called "dustbowl". Jungle clearance in South America for temporary agriculture, without subsequent re-afforestation, is also resulting in man-made deserts. The methods formerly used to control vegetation were relatively simple, consisting of ploughing and weed eradication. However, modern usage of herbicides has greatly increased the means at the disposal of man to alter or control vegetation in all stages of development and consequently the danger of major changes over wide-spread areas has also increased.

## 7. Seedling Establishment

Root growth ensuring water supply, seedling vigour in piercing the soil surface and the ability to begin photosynthesis are some of the factors which ensure the establishment, especially of the seedlings from small seeds. The fact that most small seeds have as their main storage materials fats which are of high caloric value may also be important. Also small, fat-containing dicotyledenous seeds have epigeal germination and green cotyledons.

The germinating seed must first of all establish anchorage by the root in the soil and ensure commencement of water and solute absorption.

The question of the pressure that must be developed by the rootlet to pierce the soil was discussed already by Pfeffer (1893). Some empirical data were collected over the years, but no attempt to evaluate the relevant soil parameters was made. A more precise attempt to explain certain features of growth of pea and wheat roots were made by Barley *et al.* (1965). In these experiments, soil density, matric potential and apparent soil cohesion were varied and root growth followed. Soil cores of different densities and matric potential were prepared and placed in the experimental pots to form a resistance to the advance of the roots. This resistance delayed further root growth by at least 24 hours and induced an increase in root diameter. Production of lateral roots was reduced by the resistance of the cores. Eventually the roots did penetrate the cores and grew through them. The data show that soil strength does have an important effect on the ability of the roots to penetrate into the soil, especially through clods or finely structured layers. Fine metal probes and penetrometers were developed to study the problem more precisely (Farrell and Greacen, 1966). It seems that the maximal pressure that can be developed by the growing root will be close to 10 bars, a value that was already suggested by Pfeffer (1893).

The diffusion of solutes toward the growing root in a fairly dry soil may apparently result in concentrations, at the root surface, ten times higher than the average concentration in the soil solution (Passioura and Frere, 1967). This may also create osmotic problems in water absorption.

When anchorage is accomplished and water and solute absorption is ensured there is still a problem of piercing the soil surface and raising the plumule. Natural rainfall, or irrigation, cause the breakdown and rearrangement of the structural units of the soil surface. Due to water flow and splash, and due to sedimentation of the disturbed soil particles the surface tends to be covered by a continuous layer of closely packed particles. These compact layers are often described as "seals". Besides the effect they exert in decreasing the infiltration rate and gaseous exchange of the soil, they also form a considerable mechanical impedance to seedling emergence (Arndt, 1965a, b). In some cases a single seedling is incapable of piercing the seal and only when a few seedlings germinate simultaneously are they capable of lifting the seal and breaking it, thus creating pathways for emergence above ground. Arndt (1965b) suggests a model to permit the estimation of the axial force which can be developed by the seedling in order to penetrate the seal. The seedling can be regarded as a column supporting a load. The initial load-bearing potential depends to a large extent on the cross-section at the root-stem junction. This potential decreases with the growth of the stem, or the hypocotyl, as the buckling tendency of a loaded column increases with its length, independently of the cross-sectional area. As the length of the stem increases, due to deep burial of the seed, the combined effects of the reduced cross-sectional area of the stem and increased tendency for buckling tend to reduce the effective axial force of the seedling. A shallow position of the seeds in the soil may be therefore advantageous to seedling establishment.

The seeds of the various species contain different amounts of storage materials and therefore different needs for the onset of independent photosynthesis and production of protein and other plant components. Most cotyledons are leaf-like and especially in the epigeal species they become green and apparently capable of photosynthesis.

The different epigeal species differ in the ability of their cotyledons for expansion and photosynthesis. Lovell and Moore (1970) studied the cotyledons of 11 species ranging from hypogeal, through non-expanding epigeal to species that develop a very considerable photosynthetic area. The cotyledons of the hypogeal *Pisum sativum* and *Phaseolus multiflorus* did not expand, lost weight with time and survived for a very short period. They produced almost no chlorophyll when exposed to light, had no stomata and practically no capacity for $CO_2$ fixation. *Phaseolus vulgaris*, although epigeal, has cotyledons of the hypogeal type. On the other hand, lupin, sunflower, cucumber and mustard all have expanding cotyledons, produced considerable amounts of chlorophyll, had stomata on both sides and a considerable capacity for $CO_2$ fixation.

The products of $^{14}CO_2$ fixation were not transported from the cotyledons until they reached their maximal size. Chlorophyll production in cotyledons attached to the axis was greatly enhanced compared to detached ones (Moore *et al.*, 1972). It also seems that the embryonic axis has some differential effect on the development of the photosynthetic systems in the cotyledons (Moore and Lovell, 1970). The growth rate of seedlings was highest in those species in which the cotyledons showed most expansion and the highest capacity for $CO_2$ fixation (Lovell and Moore, 1971). In such seedlings the development of the first leaves was delayed.

From the foregoing discussion of the ecology of germination it is clear that much remains to be discovered in this respect. There is an urgent need for controlled experiments and for rigorous comparisons between the ecological behaviour of seeds in the laboratory and in their natural habitat. Germination is a key process in determining plant distribution and a study of its ecology will add greatly to our understanding of plant ecology in its wider aspects.

## Bibliography

Akkerman, A. M. and Veldstra, H. (1947) *Rec. Trav. Chim. Pays-Bas* **66**, 441.
Arndt, W. (1965a) *Aust. J. Soil Res.* **3**, 45.
Arndt, W. (1965b) *Aust. J. Soil Res.* **3**, 55.
Bachthaler, G. (1957) *Z. Acker und Pflanzenbau* **103**, 128.
Ballard, L. A. T. and Grant Lipp, A. E. (1959) *Aust. J. Biol. Sci.* **12**, 342.
Barley, K., Farrell, D. A. and Greacen, E. L. (1965) *Aust. J. Soil Res.* **3**, 69.
Berg, Y. (1972) *Am. J. Bot.* **59**, 109.

Bonner, J. (1950) *Bot. Rev.* **16**, 51.
Brenchley, W. E. and Warington, K. (1933) *J. Ecol.* **21**, 103.
Brown, R. (1965) In *Handbuch der Pflanzen Physiologie* **15**, 925, Springer-Verlag.
Cannon, J. R., Corbett, N. H., Haydock, K. P., Tracey, J. G. and Webb, L. J. (1962) *Aust. J. Bot.* **10**, 119.
Capon, B. and van Asdall, W. (1967) *Ecol.* **48**, 305.
Carroll, C. R. and Janzen, D. H. (1973) *Ann. Rev. Ecol. & Syst.* **4**, 231.
Cavers, P. B. and Harper, J. L. (1966) *J. Ecol.* **54**, 367.
Cavers, P. B. and Harper, J. L. (1967) *J. Ecol.* **55**, 73.
Cohen, D. (1966) *J. Theoret. Biol.* **12**, 119.
Cohen, D. (1968) *J. Ecol.* **56**, 219.
Collis-George, N. and Hector, R. (1966) *Aust. J. Soil Res.* **4**, 145.
Come, D. (1970) *Les Obstacles à la Germination*, pp. 162, Masson & Cie, Paris.
Crocker, W. (1906) *Bot. Gaz.* **42**, 265.
Dinoor, A. (1959) M.Sc. (Agr.) Thesis, Rehovot (in Hebrew).
Duvel, J. W. T. (1905) *U.S.D.A. Bureau of Plant Industry, Bull. No. 83.*
Edwards, M. M. (1969) *J. Expt. Bot.* **20**, 876.
Edwards, W. G. H. (1972) In *Phytochemical Ecology*, p. 235 (ed. J. B. Harborne), Acad. Press, London.
Evenari, M. (1965) In *Handbuch der Pflanzen Physiologie* **16**, 691, Springer-Verlag.
Farrell, D. A. and Greacen, E. L. (1966) *Aust. J. Soil Res.* **4**, 1.
Funke, G. L. (1943) *Blume*, **5**, 281.
Gale, J., Naaman, R. and Poljakoff-Mayber, A. (1970) *Aust. J. Biol. Sci.* **23**, 947.
Gindel, L. (1960) *Nature, Lond.* **187**, 42.
Gutterman, Y., Witztum, A. and Evenari, M. (1967) *Isr. J. Bot.* **16**, 213.
Hadas, A. (1970) *Isr. J. Agr. Res.* **20**, 3.
Hadas, A. and Stibbe, E. (1973) In *Physical Aspects of Soil in Eco-systems*, p. 97 (ed. Hadas, Ed., Swartzerbruder, P., Rijtema, P. E., Fuchs, M. and Yaron, B.), Springer-Verlag, R.
Harper, J. L. and Benton, R. A. (1966) *J. Ecol.* **54**, 151.
Harper, J. L., Lovell, P. H. and Moore, K. G. (1970) *Ann. Rev. Ecol. & Syst.* **1**, 327.
Hunter, J. R. and Erickson, A. F. (1952) *Agr. J.* **44**, 107.
Juhren, M., Hiesen, W. H. and Went, F. W. (1953) *Ecology* **34**, 288.
Kendrick, R. E. and Spruit, C. J. B. (1973) *Photochem. & Photobiol.* **18**, 153.
Kidd, F. (1914) *Proc. Roy. Soc. B.* **87**, 408.
Kidd, F. (1914) *Proc. Roy. Soc. B.* **87**, 609.
Koller, D. (1954) Ph.D. Thesis, Jerusalem (in Hebrew).
Koller, D. (1955) *Bull. Res. Council, Israel* **5D**, 85.
Koller, D. (1956) *Ecology* **37**, 430.
Koller, D. (1972) In *Seed Biology*, vol. 2, p. 1. (ed. Kozlowski).
Koller, D. and Cohen, D. (1959) *Bull. Res. Council, Israel* **7D**, 175.
Koller, D. and Negbi, M. (1959) *Ecology* **40**, 20.
Koves, E. and Varga, M. (1958) *Acta Biolog. Szeged.* **4**, 13.
Kuhn, R., Jerchel, D., Moewus, F. and Moeller, E. F. (1943) *Naturwissenschaften* **31**, 468.
Lazenby, A. (1956) *Herbage Abstracts* **26**, 71.
Lerner, H. R., Mayer, A. M. and Evenari, M. (1959) *Physiol. Plant.* **12**, 245.
Lerner, H. R. and Evenari, M. (1961) *Physiol. Plant.* **14**, 229.
Lodhi, M. A. K. and Rice E. L. (1971) *Bull. Torrey Bot. Club* **98**, 83.
Lovell, P. H. and Moore, K. G. (1970) *J. Expt. Bot.* **21**, 1017.
Lovell, P. H. and Moore, K. G. (1971) *J. Expt. Bot.* **22**, 153.
Lubke, M. A. and Cavers, P. B. (1964) *Can. J. Bot.* **47**, 529.
Malcolm, C. V. (1964) *J. Royal Soc. West Australia* **47**, 73.
Mall, L. P. (1954) *Proc. Nat. Acad. Sci. India* **24**, Sec. B., 197.
Marchaim, U., Birk, Y., Dovrat, A. and Berman, T. (1972) *J. Expt. Bot.* **23**, 302.
McComb, J. A. and Andrews, R. (1974) *Aust. J. Expt. Agr. & Animal Husbandry*, **14**, 68.
Moore, K. G. and Lovell, P. H. (1970) *Planta* **93**, 284.
Moore, K. G., Bentley, K. and Lovell, P. H. (1972) *J. Expt. Bot.* **23**, 432.
Morinaga, T. (1926) *Amer. J. Bot.* **13**, 126.
Morinaga, T. (1926) *Amer. J. Bot.* **13**, 159.
Mott, J. J. (1972) *J. Ecol.* **60**, 293.
Muller, C. H. (1953) *Amer. J. Bot.* **40**, 53.
Passioura, J. B. and Frere, M. H. (1967) *Aust. J. Soil Res.* **5**, 149.

Pfeffer, W. (1893) *Abh. Sachs. Akad. Wiss.* **20**, 233.
Porter, N. G. and Wareing, P. F. (1974) *Expt. Bot.* **25**, 583.
Scherzer, R. (1954) M.Sc. Thesis, Jerusalem (in Hebrew).
Sheldon, J. C. (1974) *J. Ecol.* **62**, 47.
Shropshire, W. (1973) *Solar Energy* **15**, 99.
Shull, C. A. (1911) *Bot. Gaz.* **52**, 455.
Sifton, H. B. (1959) *Canad. J. Bot.* **37**, 719.
Simpson, G. M. (1966) *Canad. J. Bot.* **44**, 1.
Sinclair, T. R. and Lemon, R. (1973) *Solar Energy* **15**, 89.
Singh, B. N. and Lai, B. N. (1937) *J. Ind. Bot. Soc.* **16**, 129.
Soriano, A. (1953) *Rev. Invest. Agr.* **7**, 253.
Soriano, A. (1953) *Rev. Invest. Agr.* **7**, 315.
Stiles, W. (1950) *Introduction to the Principles of Plant Physiology*, Methuen, London.
Stoutjesdyk, Ph. (1972) *Acta Bot. Neerl.* **21**, 185.
Summerfield, R. J. (1973) *J. Ecol.* **61**, 387.
Sunderland, N. (1960) *J. Expt. Bot.* **11**, 236.
Thompson, P. A. (1970) *An. Bot.* **34**, 427.
Thompson, P. A. (1971a) *Physiol. Plant.* **23**, 734.
Thompson, P. A. (1971b) *J. Ecol.* **58**, 699.
Thornton, N. C. (1943–45) *Contr. Boyce, Thompson Inst.* **13**, 357.
Valio, F. M. (1973) *J. Expt. Bot.* **24**, 442.
Van Sumere, C. F., Cottenie, J., de Greef, J. and Kint, J. (1972) *Recent Adv. Phytochem.* **4**, 163.
Varga, M. (1957) *Acta Biolog. Szeged.* **3**, 213.
Varga, M. (1957) *Acta Biolog. Szeged.* **3**, 225.
Vickery, R. K. (1967) *Ecol.* **48**, 649.
Villiers, T. A. and Wareing, P. F. (1960) *Nature, Lond.* **185**, 112.
Wareing, P. F. and Foda, H. A. (1957) *Physiol. Plant.* **10**, 266.
Went, F. W. (1953) *Desert Research Proceedings, Int. Symp.* 230–240 (Jerusalem Special Publication No. 2, Res. Council, Israel).
Went, F. W. (1957) *Experimental Control of Plant Growth* 248–251, Chronica Botanica Co., Waltham, Mass.
Went, F. W., Juhren, G. and Juhren, M. C. (1952) *Ecology* **33**, 351.
Wesson, G. and Wareing, P. F. (1969) *J. Expt. Bot.* **20**, 414.
Winter, A. G. and Bublitz, W. (1953) *Naturwissenschaften* **40**, 416.
Yardeni, D. and Evenari, M. (1952) *Phyton* **2**, 11.

# INDEX OF PLANTS

179

# AUTHOR INDEX

# SUBJECT INDEX

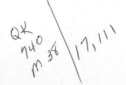